Lambert M. Surhone, Mariam T. Tennoe,
Susan F. Henssonow (Ed.)

Bombay Explosion (1944)

Lambert M. Surhone, Mariam T. Tennoe,
Susan F. Henssonow (Ed.)

Bombay Explosion (1944)

Victoria Dock, Mumbai, SS Fort Stikine, Prince Rupert, British Columbia, Fort Stikine, Colaba Observatory

Betascript Publishing

Imprint

All parts of this book are extracted from Wikipedia, the free encyclopedia (www.wikipedia.org).

You can get detailed informations about the authors of this collection of articles at the end of this book. The editors (Ed.) of this book are no authors. They have not modified or extended the original texts.

Pictures published in this book can be under different licences than the GNU Free Documentation License. You can get detailed informations about the authors and licences of pictures at the end of this book.

The content of this book was generated collaboratively by volunteers. Please be advised that nothing found here has necessarily been reviewed by people with the expertise required to provide you with complete, accurate or reliable information. Some information in this book maybe misleading or wrong. The Publisher does not guarantee the validity of the information found here. If you need specific advice (f.e. in fields of medical, legal, financial, or risk management questions) please contact a professional who is licensed or knowledgeable in that area.

Cover image: www.ingimage.com
Concerning the licence of the cover image please contact ingimage.

Contact:
VDM Publishing House Ltd.,17 Rue Meldrum, Beau Bassin,1713-01 Mauritius
Email: info@vdm-publishing-house.com
Website: www.vdm-publishing-house.com

Published in 2010
Printed in: U.S.A., U.K., Germany. This book was not produced in Mauritius.

ISBN: 978-613-1-48629-6

Contents

Articles

Bombay Explosion (1944) 1
Victoria Dock 3
Mumbai 4
SS Fort Stikine 34
Prince Rupert, British Columbia 37
Fort Stikine 46
Colaba Observatory 50
Hudson's Bay Company 53
Port Said 64
Mumbai Fire Brigade 68
Karachi 74
Maharashtra 95

References

Article Sources and Contributors 117
Image Sources, Licenses and Contributors 120

Article Licenses

License 124

Bombay Explosion (1944)

Bombay Explosion (1944)

Smoke billowing out of the site

Date	14 April 1944
Time	16:15 IST (10:45 UTC)
Location	Victoria Dock, Bombay, British India
Casualties	
800 dead	
3,000 injured	

The **Bombay Explosion** (or Bombay Docks Explosion) occurred on 14 April 1944, in the Victoria Dock of Bombay (now Mumbai) when the SS *Fort Stikine* carrying a mixed cargo of cotton bales, gold, ammunition including around 1,400 tons of explosive caught fire and was destroyed in two giant blasts, scattering debris, sinking surrounding ships and killing around 800 people.

The vessel, the voyage and cargo

The SS *Fort Stikine* was a 7142 gross ton freighter built in 1942 in Prince Rupert, British Columbia, under a lend-lease agreement, and was named for Fort Stikine, a former outpost of the Hudson's Bay Company located at what is now Wrangell, Alaska.

SS *Fort Stikine*

Sailing from Birkenhead on 24 February via Gibraltar, Port Said and Karachi, she arrived at Bombay on 12 April.

The ship carried

- explosives
- munitions
- Spitfires
- raw cotton bales
- oil barrels
- timber
- scrap iron
- gold bullion in 12.73 kg bars valued at £1–2 million.

One officer described the cargo as "just about everything that will either burn or blow up". The vessel berthed and was still awaiting unloading on 14 April.

Incident

The explosion

In the mid-afternoon around 14.00, the crew were alerted to a fire onboard. Burning somewhere in the No. 2 hold, the crew, dockside fire teams and fireboats were unable to extinguish the conflagration, despite pumping over 900 tons of water into the ship, or find the source due to the dense smoke.

At 15:50 the order to abandon ship was given, and sixteen minutes later there was a great explosion, cutting the ship in two and breaking windows over 12 km away. The two explosions were powerful enough to be recorded by seismographs at the Colaba Observatory in the city. Around two square miles were ablaze in an 800-metre arc around the ship, eleven neighbouring vessels were sunk or sinking, and the emergency personnel at the site suffered heavy losses. Attempts to fight the fire were dealt a further blow when a second explosion from the ship swept the area at 16:34.

Aftermath

Aftermath of the explosion at the harbour

It took three days to bring the fire under control, and later 8,000 men toiled for seven months to remove around 500,000 tons of debris and bring the docks back into action. The official death toll was 740, including 476 military personnel, with around 1,800 people injured; unofficial tallies run much higher. In total, twenty-seven other vessels were sunk or damaged in both Victoria dock and the neighbouring Prince's Dock.

Many families lost all their belongings and were left with just the clothes on their back. The government took full responsibility for the disaster and monetary compensation was paid to citizens who made a claim for loss or damage to property.

A piece of propeller that landed in St. Xaviers High School, some three kilometres from the docks.

During normal dredging operations carried out periodically to maintain the depth of the docking bays one or two gold bars were found intact sporadically as late as the 1970s and returned to the British government. Once in every few years, gold bricks are recovered from Mumbai harbour, reminding everyone of the great tragedy, even six decades after the incident. Mumbai Fire Brigade's headquarters at Byculla has a memorial built in the memory of numerous fire fighters who died during this explosion. Fire Safety Week is observed all over Maharashtra from 14 April to 21 April in memory of Fire fighters who died in this explosion

See

Mumbai Fire Brigade

External links

- The Great Bombay Explosion by John Ennis [1]
- Ship Explosions: SS Fort Stikine, Page 7-10 [2]
- The Great Bombay Explosion [3] By Lawrence Wilson
- Bombay Harbour Explosion [4]
- The day it rained gold [5]
- Ships lost in the Bombay Explosion [6]

Geographical coordinates: 18°57′10″N 72°50′42″E

References

[1] http://members.tripod.com/~merchantships/fortcrevierepilogue.html
[2] http://www.petervogel.us/lastwave/chapters/LastWave_Ch7.pdf
[3] http://www.holysmoke.org/great-bombay.htm
[4] http://www.merchantnavyofficers.com/bomEx.html#memorial
[5] http://www.tribuneindia.com/2000/20000727/edit.htm#5
[6] http://members.tripod.com/~merchantships/fortcreviershipslost.html

Victoria Dock

Victoria Dock may refere to:

United Kingdom

- Royal Victoria Dock, London
- Victoria Dock, Liverpool
- Victoria Dock in Kingston upon Hull

Australia

- Victoria Dock (Hobart)
- Victoria Dock (Melbourne) - previously part of the Port of Melbourne, now part of the Melbourne Docklands redevelopment and called Victoria Harbour

South Africa

- Victoria & Alfred Waterfront, Cape Town

Mumbai

Geographical coordinates: 18°N 73°E

Mumbai (मुंबई)
Bombay

— **Metropolitan city** —

Clockwise from top: Skyline at Cuffe Parade, the Rajabai Clock Tower, the Taj Mahal Hotel, Nariman Point and Gateway of India

Mumbai (मुंबई)

Location of Mumbai (मुंबई)
in Maharashtra and India

Coordinates	18°58′30″N 72°49′33″E
Country	India
State	Maharashtra
District(s)	Mumbai City Mumbai Suburban
Municipal commissioner	Swadhin Kshatriya
Mayor	Shraddha Jadhav
Population • Density • Metro	13,830,884[1] (1st) (2010) • 22,922 /km^2 (/sq mi) • 21,900,967[2] (1st) (2010)

Time zone	IST (UTC+5:30)
Area • Elevation	603.4 km^2 (233 sq mi) • 14 m (46 ft)
Website	www.mcgm.gov.in [3]

Mumbai (Marathi: मुंबई, *Mumbaī*, IPA: [ˈmʊm.bəi] Wikipedia:Media helpFile:Mumbai_pronunciation.ogg), formerly called **Bombay**, is the capital of the Indian state of Maharashtra. Mumbai, the most populous city in India, is the second most populous city in the world, with a population of approximately 14 million.[1] Along with the neighbouring urban areas, including the cities of Navi Mumbai and Thane, it is one of the most populous urban regions in the world.[4] Mumbai lies on the west coast of India and has a deep natural harbour. As of 2009, Mumbai was named an Alpha world city.[5] Mumbai is also the richest city in India.[6]

The seven islands that came to constitute Bombay were home to communities of fishing colonies. For centuries, the islands came under the control of successive indigenous empires before being ceded to the Portuguese and subsequently to the British East India Company. During the mid-18th century, Bombay was reshaped by the British with large-scale civil engineering projects,[7] and emerged as a significant trading town. Economic and educational development characterised the city during the 19th century. It became a strong base for the Indian independence movement during the early 20th century. When India became independent in 1947, the city was incorporated into Bombay State. In 1960, following the Samyukta Maharashtra movement, a new state of Maharashtra was created with Bombay as capital. It was renamed Mumbai in 1995.[8]

Mumbai is the commercial and entertainment centre of India, generating 5% of India's GDP,[9] and accounting for 25% of industrial output, 40% of maritime trade, and 70% of capital transactions to India's economy.[10] Mumbai is home to important financial institutions such as the Reserve Bank of India, the Bombay Stock Exchange, the National Stock Exchange of India and the corporate headquarters of numerous Indian companies and multinational corporations. The city also houses India's Hindi film and television industry, known as Bollywood. Mumbai's business opportunities, as well as its potential to offer a higher standard of living, attract migrants from all over India and, in turn, make the city a potpourri of many communities and cultures.

Toponymy

The name *Mumbai* is an eponym, etymologically derived from *Mumba* or *Maha-Amba*—the name of the Koli goddess Mumbadevi—and *Aai*, "mother" in Marathi.[11] The former name *Bombay* had its origins in the 16th century when the Portuguese arrived in the area and called it by various names, which finally took the written form *Bombaim*, still common in current Portuguese use.[12] After the British gained possession of the city in the 17th century, it was believed to be anglicised to *Bombay* from the Portuguese *Bombaim*.[13] The city was known as *Mumbai* or *Mambai* to Marathi speakers and as *Bambai* in Hindi, Persian and Urdu. It is sometimes still referred to by its older names, such as *Kakamuchee* and *Galajunkja*.[14] [15] The English name was officially changed to its Marathi pronunciation of *Mumbai* in November 1995.[16] This came at the insistence of the Hindu nationalist Shiv Sena party, that had just won the Maharashtra state elections and mirrored similar name changes across the country. However, the city is still commonly referred to as Bombay by many of its residents and Indians from other regions as well.[17]

The temple of local Hindu goddess Mumbadevi, after which the city of Mumbai derives its name

A widespread explanation of the origin of the traditional English name *Bombay* holds that it was derived from a Portuguese name meaning "good bay". This is based on the fact that *bom* (masc.) is Portuguese for "good" whereas the English word "bay" is similar to the Portuguese *baía* (fem., *bahia* in old spelling). The normal Portuguese rendering of "good bay" would have been *boa bahia* rather than the grammatically incorrect *bom bahia*. However, it is possible to find the form *baim* (masc.) for "little bay" in 16th-century Portuguese.[18] Portuguese scholar José Pedro Machado in his *Dicionário Onomástico Etimológico da Língua Portuguesa* (Portuguese Dictionary of Onomastics and Etymology), seems to reject the "Bom Bahia" hypothesis, asserting that Portuguese records mentioning the presence of a bay at the place led the English to assume that the noun (*bahia*, "bay") was an integral part of the Portuguese toponym, hence the English version Bombay, adapted from Portuguese.[19]

Mirat-i-Ahmedi referred to the city as *Manbai* in 1507.[20] The earliest Portuguese writer to refer to the city as *Bombaim* was Gaspar Correia in 1508, as recorded in his *Lendas da Índia* ("Legends of India").[21] [22] Portuguese explorer Duarte Barbosa mentions a reference to the city in a complex form, as *Tana-Maiambu* or *Benamajambu* in 1516.*Tana* appears to refer to the name of the adjoining town of Thane, and *Maiambu* seems to refer to Mumba-Devi, the Hindu goddess after which the place is named in Marathi.[23] Other variations of the name recorded in the 16th and the 17th centuries are, *Mombayn* (1525), *Bombay* (1538), *Bombain* (1552), *Bombaym* (1552), *Monbaym* (1554), *Mombaim* (1563), *Mombaym* (1644), *Bambaye* (1666), *Bombaiim* (1666), *Bombeye* (1676), and *Boon Bay* (1690).[12] [24]

History

Early history

Kanheri Caves served as a centre of Buddhism in Western India during ancient times

Mumbai is built on what was once an archipelago of seven islands: Bombay Island, Parel, Mazagaon, Mahim, Colaba, Worli, and Old Woman's Island (also known as *Little Colaba*).[25] Pleistocene sediments found along the coastal areas around Kandivali in northern Mumbai by archaeologist Todd in 1939 suggest that these islands were inhabited since the Stone Age.[26] It is not exactly known when these islands were first inhabited. Perhaps at the beginning of the Common era (2000 years ago), or even possibly earlier, they came to be occupied by the Koli fishing community.[27] In the third century BCE, the islands formed part of the Maurya Empire, during its expansion in the south, ruled by the Buddhist emperor, Ashoka of Magadha.[28] The Kanheri Caves in Borivali were excavated in the mid-third century BCE,[29] and served as an important centre of Buddhism in Western India during ancient times.[30] The city then was known as *Heptanesia* (Ancient Greek: A Cluster of Seven Islands) to the Greek geographer Ptolemy in 150 CE.[31]

Between the second century BCE and ninth century CE, the islands came under the control of successive indigenous dynasties: Satavahanas, Western Kshatrapas, Abhiras, Vakatakas, Kalachuris, Konkan Mauryas, Chalukyas and Rashtrakutas,[32] before being ruled by the Silhara dynasty from 810 to 1260.[33]

Some of the oldest edifices in the city built during this period are, Jogeshwari Caves (between 520 to 525),[34] Elephanta Caves (between the sixth to seventh century),[35] Walkeshwar Temple (10th century),[36] and Banganga

Tank (12th century).[37] King Bhimdev founded his kingdom in the region in the late 13th century, and established his capital in *Mahikawati* (present day Mahim).[38] The Pathare Prabhus, one of the earliest known settlers of the city, were brought to *Mahikawati* from Saurashtra in Gujarat around 1298 by Bhimdev.[39] The Delhi Sultanate annexed the islands in 1347–48, and controlled it till 1407. During this time, the islands were administered by the Muslim Governors of Gujarat, who were appointed by the Delhi Sultanate.[40] [41] The islands were later governed by the independent Gujarat Sultanate, which was established in 1407. The Sultanate's patronage led to the construction of many mosques, prominent being the Haji Ali Dargah in Worli, built in honour of the Muslim saint Haji Ali in 1431.[42] From 1429 to 1431, the islands were a source of contention between the Gujarat Sultanate and the Bahamani Sultanate of Deccan.[43] [44] In 1493, Bahadur Khan Gilani of the Bahamani Sultanate attempted to conquer the islands, but was defeated.[45]

European rule

The Haji Ali Dargarh was built in 1431, when Mumbai was under the Gujarat Sultanate

The Mughal Empire, founded in 1526, was the dominant power in the Indian subcontinent during the mid-16th century.[46] Growing apprehensive of the power of the Mughal emperor Humayun, Sultan Bahadur Shah of the Gujarat Sultanate was obliged to sign the Treaty of Bassein with the Portuguese Empire on 23 December 1534. According to the treaty, the seven islands of Bombay, the nearby strategic town of Bassein and its dependencies were offered to the Portuguese. The territories were later surrendered on 25 October 1535.[47] The Portuguese were actively involved in the foundation and growth of their Roman Catholic religious orders in Bombay.[48] Some of the oldest Catholic churches in the city such as the St. Michael's Church at Mahim (1534),[49] St. John the Baptist Church at Andheri (1579),[50] St. Andrew's Church at Bandra (1580),[51] and Gloria Church at Byculla (1632),[52] date from the Portuguese era. On 11 May 1661, the marriage treaty of Charles II of England and Catherine of Braganza, daughter of King John IV of Portugal, placed the islands in possession of the British Empire, as part of Catherine's dowry to Charles.[53] However, Salsette, Bassein, Mazagaon, Parel, Worli, Sion, Dharavi, and Wadala still remained under Portuguese possession. From 1665 to 1666, the British managed to acquire Mahim, Sion, Dharavi, and Wadala.[54]

These islands were in turn leased to the British East India Company in 1668 for a sum of £10 per annum by the Royal Charter of 27 March 1668.[55] The population quickly rose from 10,000 in 1661, to 60,000 in 1675.[56] The islands were subsequently attacked by Yakut Khan, the Siddi admiral of the Mughal Empire, in October 1672,[57] Rickloffe van Goen, the Governor-General of Dutch India on 20 February 1673,[58] and Siddi admiral Sambal on 10 October 1673.[57] In 1687, the British East India Company transferred its headquarters from Surat to Bombay. The city eventually became the headquarters of the Bombay Presidency.[59] Following the transfer, Bombay was placed at the head of all the Company's establishments in India.[60] Towards the end of the 17th century, the islands again suffered incursions from Yakut Khan in 1689–90.[61] The Portuguese presence ended in Bombay when the Marathas under *Peshwa* Baji Rao I captured Salsette in 1737, and Bassein in 1739.[62] By the middle of the 18th century, Bombay began to grow into a major trading town, and received a huge influx of migrants from across India.[63] Later, the British occupied Salsette on 28 December 1774. With the Treaty of Surat (1775), the British formally gained control of Salsette and Bassein, resulting in the First Anglo-Maratha War.[64] The British were able to secure Salsette from the Marathas without violence through the Treaty of Purandar (1776),[65] and later through the Treaty of Salbai (1782), signed to settle the outcome of the First Anglo-Maratha War.[66]

Ships in Bombay Harbour (c. 1731). Bombay emerged as a significant trading town during the mid-18th century

From 1782 onwards, the city was reshaped with large-scale civil engineering projects aimed at merging all the seven islands into a single amalgamated mass. This project, known as Hornby Vellard, was completed by 1784.[7] In 1817, the British East India Company under Mountstuart Elphinstone defeated Baji Rao II, the last of the Maratha *Peshwa* in the Battle of Kirkee.[67] Following his defeat, almost the whole of the Deccan came under British suzerainty, and were incorporated in Bombay Presidency. The success of the British campaign in the Deccan witnessed the freedom of Bombay from all attacks by native powers.[68] By 1845, the seven islands were coalesced into a single landmass by the Hornby Vellard project.[69] On 16 April 1853, India's first passenger railway line was established, connecting Bombay to the neighbouring town of Thane.[70] During the American Civil War (1861–1865), the city became the world's chief cotton trading market, resulting in a boom in the economy that subsequently enhanced the city's stature.[71] The opening of the Suez Canal in 1869 transformed Bombay into one of the largest seaports on the Arabian Sea.[72] In September 1896, Bombay was hit by a bubonic plague epidemic where the death toll was estimated at 1,900 people per week.[73] About 850,000 people fled Bombay and the textile industry was adversely affected.[74] As the capital of the Bombay Presidency, it witnessed the Indian independence movement, with the Quit India Movement in 1942 and the The Royal Indian Navy Mutiny in 1946 being its most notable events.[75] [76]

Independent India

The *Hutatma Chowk* memorial, built to honour the martyrs of the Samyukta Maharashtra movement. (Flora Fountain is on its left in the background)

After India's independence in 1947, the territory of the Bombay Presidency retained by India was restructured into Bombay State. The area of Bombay State increased, after several erstwhile princely states that joined the Indian union were integrated into Bombay State. Subsequently, the city became the capital of Bombay State.[77] On April 1950, Municipal limits of Bombay were expanded by merging the Bombay Suburban District and Bombay City to form Greater Bombay Municipal Corporation.[78] The Samyukta Maharashtra movement to create a separate Maharashtra state including Bombay was at its height in the 1950s. In the *Lok Sabha* discussions in 1955, the Congress party demanded that the city be constituted as an autonomous city-state.[79] The States Reorganisation Committee recommended a bilingual state for Maharashtra–Gujarat with Bombay as its capital in its 1955 report. Bombay Citizens' Committee, an advocacy group comprising of leading Gujarati industrialists lobbied for Bombay's independent status.[80] Following protests during the movement in which 105 people were killed by police, Bombay State was reorganised on linguistic lines on 1 May 1960.[81] [82] Gujarati-speaking areas of Bombay State were partitioned into the state of Gujarat.[83] Maharashtra State with Bombay as its capital was formed with the merger of Marathi-speaking areas of Bombay State, eight districts from Central Provinces and Berar, five districts from Hyderabad State, and numerous princely states enclosed between them.[84] As a memorial to the martyrs of the Samyukta Maharashtra movement, Flora Fountain was renamed as *Hutatma Chowk* (Martyr's Square), and a memorial was erected.[85]

The following decades saw massive expansion of the city and its suburbs. In the late 1960s, Nariman Point and Cuffe Parade were reclaimed and developed.[86] The Bombay Metropolitan Region Development Authority (BMRDA) was set up on 26 January 1975 by the Government of Maharashtra as an apex body for planning and co-ordination of development activities in the Bombay metropolitan region.[87] In August 1979, a sister township of New Bombay was founded by City and Industrial Development Corporation (CIDCO) across Thane and Raigad districts to help the dispersal and control of Bombay's population.[88] Textile industry in Bombay largely disappeared after the massive 1982 Great Bombay Textile Strike, in which nearly 250,000 workers in more than 50 textile mills went on strike.[89] The Jawaharlal Nehru Port, which currently handles 55–60% of India's containerized cargo, was commissioned on 26 May 1989 at Nhava Sheva with a view to de-congest Bombay Harbour and to serve as a hub port for the city.[90] The geographical limits of Greater Bombay were coextensive with municipal limits of Greater Bombay. On 1 October 1990, the Greater Bombay district was bifurcated to form 2 revenue districts namely, Bombay City and Bombay Suburban, though they were administered by same Municipal Administration.[91]

The past two decades have seen an increase in violence in the hitherto largely peaceful city. Following the demolition of the Babri Masjid in Ayodhya, the city was rocked by the Hindu-Muslim riots of 1992–93 in which more than 1,000 people were killed.[92] On 12 March 1993, a series of 13 co-ordinated bombings at several city landmarks by Islamic extremists and the Bombay underworld resulted in 257 deaths and over 700 injuries.[93] In 2006, 209 people were killed and over 700 injured when seven bombs exploded on the city's commuter trains.[94] In 2008, a series of ten coordinated attacks by armed terrorists for three days resulted in 173 deaths, 308 injuries, and severe damage to a couple of heritage landmarks and prestigious hotels.[95] Today, Mumbai is the commercial capital of India and has evolved into a global financial hub.[96] For several decades it has been the home of India's main financial services, and a focus for both infrastructure development and private investment.[97] From being an ancient fishing community and a colonial centre of trade, Mumbai has become South Asia's largest city and home of the world's most prolific film industry.[98]

Geography

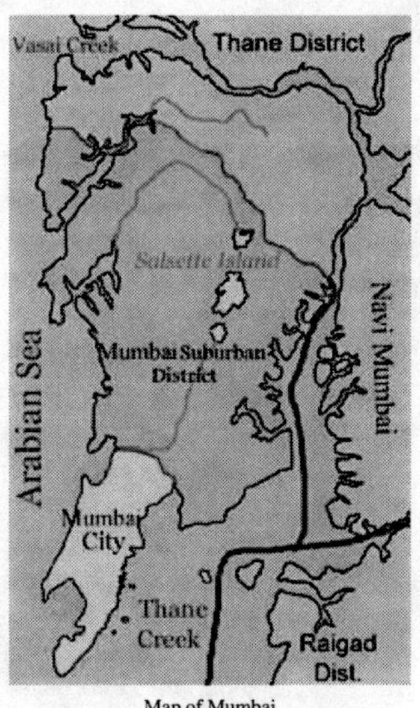

Map of Mumbai

Mumbai consists of two distinct regions: Mumbai City district and Mumbai Suburban district, which form two separate revenue districts of Maharashtra.[99] The city region is also commonly referred to as the *Island City* or South Mumbai.[100] The total area of Mumbai is 603.4 km^2 (233 sq mi),[101] with the area of 437.71 km^2 (169 sq mi), the island city spanning 67.79 km^2 (26 sq mi) and the suburban district spanning 370 km^2 (143 sq mi), coming under the administration of Brihanmumbai Municipal Corporation (BMC), while remaining area belongs to Defence, Mumbai Port Trust, Atomic Energy Commission and Borivali National Park, which are out of the jurisdiction of the BMC.[102]

Mumbai lies at the mouth of the Ulhas River on the western coast of India, in the coastal region known as the Konkan. It sits on Salsette Island, partially shared with the Thane district.[103] Mumbai is surrounded by the Arabian Sea to the west.[104] Many parts of the city lie just above sea level, with elevations ranging from 10 m (33 ft) to 15 m (49 ft);[105] the city has an average elevation of 14 m (46 ft).[106] Northern Mumbai (Salsette) is hilly,[107] and the highest point in the city is 450 m (1476 ft) at Salsette in the Powai-Kanheri ranges.[108] Sanjay Gandhi National Park (Borivali National Park) is located partly in the Mumbai suburban district, and partly in the Thane district, and it extends over an area of 103.09 km^2 (39.80 sq mi).[109]

Apart from the Bhatsa Dam, there are six major lakes that supply water to the city: Vihar, Lower Vaitarna, Upper Vaitarna, Tulsi, Tansa and Powai.[110] Tulsi Lake and Vihar Lake are located in Borivili National Park, within the city's limits.[111] The supply from Powai lake, also within the city limits, is used only for agricultural and industrial purposes.[112] Three small rivers, the Dahisar River, Poinsar (or Poisar) and Ohiwara (or Oshiwara) originate within the park, while the polluted Mithi River originates from Tulsi Lake and gathers water overflowing from Vihar and Powai Lakes.[113] The coastline of the city is indented with numerous creeks and bays, stretching from Thane creek on the eastern to Madh Marve on the western front.[114] The eastern coast of Salsette Island is covered with large mangrove swamps, rich in biodiversity, while the western coast is mostly sandy and rocky.[115]

Soil cover in the city region is predominantly sandy due to its proximity to the sea. In the suburbs, the soil cover is largely alluvial and loamy.[116] The underlying rock of the region is composed of black Deccan basalt flows, and their acidic and basic variants dating back to the late Cretaceous and early Eocene eras.[117] Mumbai sits on a seismically active zone owing to the presence of 23 fault lines in the vicinity.[118] The area is classified as a Seismic Zone III region,[119] which means an earthquake of up to magnitude 6.5 on the Richter-scale may be expected.[120]

Climate

Mumbai has a tropical climate, specifically a tropical wet and dry climate under the Köppen climate classification, with seven months of dryness and peak of rains in July.[121] The cold season from December to February is followed by the summer season from March to June. The period from June to about the end of September constitutes the south-west monsoon season, and October and November form the post-monsoon season.[122] Between June and September, the south west monsoon rains lash the city. Pre-monsoon showers are received in May. Occasionally, north-east monsoon showers occur in October and November. The maximum annual rainfall ever recorded was 3452 millimetres (135.9 in) for 1954.[123] The highest rainfall recorded in a single day was 944 millimetres (37.17 in) on 26 July 2005.[124] The average total annual rainfall is 2146.6 millimetres (84.51 in) for the Island City, and 2457 millimetres (96.73 in) for the suburbs.[123]

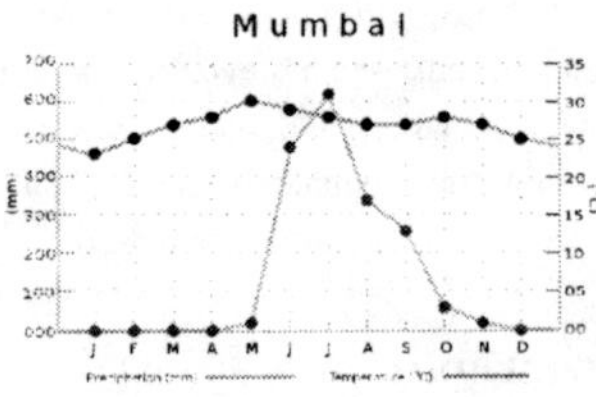

Average temperature and precipitation in Mumbai

The average annual temperature is 27.2 °C (81.0 °F), and the average annual precipitation is 216.7 centimetres (85.31 in).[125] In the Island City, the average maximum temperature is 31.2 °C (88.2 °F), while the average minimum temperature is 23.7 °C (74.7 °F). In the suburbs, the daily mean maximum temperature range from 29.1 °C (84.4 °F) to 33.3 °C (91.9 °F), while the daily mean minimum temperature ranges from 16.3 °C (61.3 °F) to 26.2 °C (79.2 °F).[123] The record high is 40.2 °C (104.4 °F) on 28 March 1982,[126] and the record low is 7.4 °C (45.3 °F) on 27 January 1962.[127]

Climate data for Mumbai

Month	Jan	Feb	Mar	Apr	May	Jun	Jul	Aug	Sep	Oct	Nov	Dec	Year
Average high °C (°F)	31 (88)	31 (88)	33 (91)	33 (91)	33 (91)	32 (90)	30 (86)	29 (84)	30 (86)	33 (91)	33 (91)	32 (90)	31 (88)
Average low °C (°F)	16 (61)	17 (63)	21 (70)	24 (75)	26 (79)	26 (79)	25 (77)	25 (77)	24 (75)	23 (73)	21 (70)	18 (64)	21 (70)
Precipitation mm (inches)	0.6 (0.02)	1.5 (0.06)	0.1 (0)	0.6 (0.02)	13 (0.51)	574 (22.6)	868 (34.17)	553 (21.77)	306 (12.05)	63 (2.48)	15 (0.59)	5.6 (0.22)	2400.4 (94.5)
Source: wunderground.com [128] *27 November 2008*													

Economy

The Bombay Stock Exchange is the oldest stock exchange in Asia

Mumbai is India's largest city and is considered the financial capital of the country as it generates 5% of the total GDP.[9] [96] It serves as an economic hub of India, contributing 10% of factory employment, 25% of industrial output, 33% of income tax collections, 60% of customs duty collections, 20% of central excise tax collections, 40% of India's foreign trade and Rs. 40 billion (US$ 890 million) in corporate taxes.[129] Mumbai's GDP is Rs 200483 crore (US$ 44.71 billion),[130] and its per-capita income is Rs. 128000 (US$ 2850)[6] , which is almost three times the national average.[69] Many of India's numerous conglomerates (including Larsen and Toubro, State Bank of India, Life Insurance Corporation of India, Tata Group, Godrej and Reliance),[96] and five of the Fortune Global 500 companies are based in Mumbai.[131] Many foreign banks and financial institutions also have branches in this area,[96] with the World Trade Centre being the most prominent one.[132] Until the 1970s, Mumbai owed its prosperity largely to textile mills and the seaport, but the local economy has since been diversified to include engineering, diamond-polishing, healthcare and information technology.[133] As of 2008, the Globalization and World Cities Study Group (GaWC) has ranked Mumbai as an "Alpha world city", third in its categories of Global cities.[134] Mumbai is the 4th most expensive office market in the world.[135]

India's 300 million strong middle-class population is growing at an annual rate of 5%.[136] Shown here is an upmarket residential area of Mumbai.

State and central government employees make up a large percentage of the city's workforce. Mumbai also has a large unskilled and semi-skilled self employed population, who primarily earn their livelihood as hawkers, taxi drivers, mechanics and other such blue collar professions. The port and shipping industry is well established, with Mumbai Port being one of the oldest and most significant ports in India.[137] In Dharavi, in central Mumbai, there is an increasingly large recycling industry, processing recyclable waste from other parts of the city; the district has an estimated 15,000 single-room factories.[138]

Most of India's major television and satellite networks, as well as its major publishing houses, are headquartered in Mumbai. The centre of the Hindi movie industry, Bollywood, is the largest film producer in India and one of the largest in the world as well as centre of Marathi Film Industry.[139] [140] Along with the rest of India, Mumbai, its commercial capital, has witnessed an economic boom since the liberalisation of 1991, the finance boom in the mid-nineties and the IT, export, services and outsourcing boom in 2000s.[141] Mumbai has been ranked 48th on the Worldwide Centres of Commerce Index 2008.[142] In April 2008, Mumbai was ranked seventh in the list of "Top Ten Cities for Billionaires" by *Forbes* magazine,[143] and first in terms of those billionaires' average wealth.[144]

Civic administration

The Bombay High Court exercises jurisdiction over Maharashtra, Goa, Daman and Diu, and Dadra and Nagar Haveli

Mumbai, extending from Colaba in the south, to Mulund and Dahisar in the north, and Mankhurd in the east, is administered by the Brihanmumbai Municipal Corporation (BMC).[104] The BMC is in charge of the civic and infrastructure needs of the metropolis.[145] The Mayor is usually chosen through indirect election by the councillors from among themselves for a term of two and half years. The Municipal Commissioner is the chief Executive Officer and head of the executive arm of the Municipal Corporation. All executive powers are vested in the Municipal Commissioner who is an Indian Administrative Service (IAS) officer appointed by the state government. Although the Municipal Corporation is the legislative body that lays down policies for the governance of the city, it is the Commissioner who is responsible for the execution of the policies. The Commissioner is appointed for a fixed term as defined by state statute. The powers of the Commissioner are those provided by statute and those delegated by the Corporation or the Standing Committee.[146]

The two revenue districts of Mumbai come under the jurisdiction of a District Collector.[147] The Collectors are in charge of property records and revenue collection for the Central Government, and oversee the national elections held in the city.[148]

The Mumbai Police is headed by a Police Commissioner, who is an Indian Police Service (IPS) officer. The Mumbai Police comes under the state Home Ministry.[149] The city is divided into seven police zones and seventeen traffic police zones,[102] each headed by a Deputy Commissioner of Police.[150] The Traffic Police is a semi-autonomous body under the Mumbai Police. The Mumbai Fire Brigade department is headed by the Chief Fire Officer, who is assisted by four Deputy Chief Fire Officers and six Divisional Officers.[102]

Mumbai is the seat of the Bombay High Court, which exercises jurisdiction over the states of Maharashtra and Goa, and the Union Territories of Daman and Diu and Dadra and Nagar Haveli.[151] Mumbai also has two lower courts, the Small Causes Court for civil matters, and the Sessions Court for criminal cases.[152] Mumbai also has a special TADA (Terrorist and Disruptive Activities) court for people accused of conspiring and abetting acts of terrorism in the city.[153]

Politics

First session of the Indian National Congress in Bombay (28–31 December 1885)

Mumbai has been a traditional stronghold and birthplace of the Indian National Congress, also known as the Congress Party.[154] The first session of the Indian National Congress was held in Bombay from 28–31 December 1885.[155] The city played host to the Indian National Congress six times during its first 50 years, and became a strong base for the Indian independence movement during the 20th century.[156] The 1960s saw the rise of regionalist politics in Bombay, with the formation of the Shiv Sena on 19 June 1966, out of a feeling of resentment about the relative marginalisation of the native Marathi people in Maharashtra.[157] The party headed a campaign to expel South Indian and North Indian migrants by force.[158] The Congress had dominated the politics of Bombay from independence until the early 1980s, when the Shiv Sena won the 1985

Bombay municipal corporation elections.[159] In 1989, the Bharatiya Janata Party (BJP), a major national political party, forged an electoral alliance with the Shiv Sena to dislodge the Congress in the Maharashtra Legislative Assembly elections. In 1999, the Nationalist Congress Party (NCP) separated from the Congress, but later allied with the Congress, to form a joint venture known as the Democratic Front.[160] Currently, other parties such as Maharashtra Navnirman Sena (MNS), Samajwadi Party (SP), Bahujan Samaj Party (BSP), and several independent candidates also contest elections in the city.[161]

In the Indian national elections held every five years, Mumbai is represented by six parliamentary constituencies: Mumbai North, Mumbai North West, Mumbai North East, Mumbai North Central, Mumbai South Central, and Mumbai South.[162] A Member of Parliament (MP) to the *Lok Sabha*, the lower house of the Indian Parliament, is elected from each of the parliamentary constituencies. In the 2009 national elections, out of the six parliamentary constituencies, five were won by the Congress, and one by the NCP.[163] In the Maharashtra state assembly elections held every five years, Mumbai is represented by 36 assembly constituencies.[164] [165] A Member of the Legislative Assembly (MLA) to the Maharashtra *Vidhan Sabha* (Legislative Assembly) is elected from each of the assembly constituencies. In the 2009 state assembly elections, out of the 36 assembly constituencies, 17 were won by the Congress, 6 by the MNS, 5 by the BJP, 4 by the Shiv Sena, and 1 by SP.[166] Elections are also held every five years to elect corporators to power in the BMC.[167] The Corporation comprises 227 directly elected Councillors representing the 24 municipal wards, five nominated Councillors having special knowledge or experience in municipal administration, and a Mayor whose role is mostly ceremonial.[168] [169] [170] In the 2007 municipal corporation elections, out of the 227 seats, the Shiv Sena-BJP alliance secured 111 seats, holding power in the BMC, while the Congress-NCP alliance bagged 85 seats.[171] The tenure of the Mayor, Deputy Mayor, and Municipal Commissioner is two and a half years.[172]

Transport

Public Transport

Public transport systems in Mumbai include the Mumbai Suburban Railway, Brihanmumbai Electric Supply and Transport (BEST) buses, black and yellow metered taxis, auto rickshaws and ferries. Suburban railway and BEST bus services together accounted for about 88% of the passenger traffic in 2008.[173] Auto rickshaws are allowed to operate only in the suburban areas of Mumbai, while taxis are allowed to operate throughout Mumbai, but generally operate in South Mumbai.[174] Taxis and rickshaws in Mumbai are required by law to run on Compressed Natural Gas,[175] and are a convenient, economical, and easily available means of transport.[174] Mumbai had about 1.53 million vehicles in 2008,[176] 56,459 black and yellow taxis, and 1,02,224 auto rickshaws, as of 2005.[177]

Road

Mumbai is served by National Highway 3, National Highway 4 and National Highway 8 of India's National Highways system.[178] The Mumbai-Pune Expressway was the first expressway built in India,[179] while the Mumbai-Vadodara Expressway,[180] Western Freeway and Eastern Freeway is under construction. The Bandra-Worli Sea Link bridge, along with Mahim Causeway, links the island city to the western suburbs.[181] The three major road arteries of the city are the Eastern Express Highway from Sion to Thane, the Sion Panvel Expressway from Sion to Panvel and the Western Express Highway from Bandra to Borivali.[182]

A BEST Starbus. BEST buses carry a total of 4.5 million passengers daily

Mumbai's bus services carried over 5.5 million passengers per day in 2008.[173] Public buses run by BEST cover almost all parts of the metropolis, as well as parts of Navi Mumbai, Mira-Bhayandar and Thane.[183] Buses are generally favored for commuting short to medium distances, while train fares are more economical for longer distance commutes.[184] The BEST operates a total of 4,013 buses[185] with CCTV Camera installed,[186] ferrying 4.5 million passengers daily[173] over 390 routes.[187] Its fleet consists of single-decker, double-decker, vestibule, low-floor, disabled-friendly, air-conditioned and Euro III compliant Compressed Natural Gas powered buses.[185] Maharashtra State Road Transport Corporation (MSRTC) buses provide intercity transport and connect Mumbai with other major cities of Maharashtra and India.[188] [189] Navi Mumbai Municipal Transport (NMMT) also operate its Volvo buses in Mumbai, from Navi Mumbai to Bandra, Dindoshi and Borivali.[190] The Mumbai Darshan is a tourist bus service which explores numerous tourist attractions in Mumbai.[191] Mumbai BRTS (Bus Rapid Transit System) lanes have been planned throughout Mumbai, with buses running on seven routes as of March 2009.[192] Though 88% of the city's commuters travel by public transport, Mumbai still continues to struggle with traffic congestion.[193] Mumbai's transport system has been categorized as one of the most congested in the world.[194]

Due to further aggravation of congestion on roads due to hawkers and parked vehicles, MMRDA has initiated the Mumbai Skywalks project to provide quick and safe pedestrian dispersal from highly congested areas such as Mumbai Suburban Railway stations to heavily targeted destinations.[195]

Rail

Mumbai is the headquarters of two of Indian Railways' zones: the Central Railway (CR) headquartered at Chhatrapati Shivaji Terminus (formerly Victoria Terminus), and the Western Railway (WR) headquartered at Churchgate.[196] The backbone of the city's transport, the Mumbai Suburban Railway, consists of three separate rail networks: Central, Western, and Harbour Line, running the length of the city, in the north-south direction.[197] Mumbai's suburban rail systems carried a total of 6.3 million passengers every day in 2007,[198] which is more than half of the Indian Railways daily carrying capacity. Trains are overcrowded during peak hours, with nine-car trains of rated capacity 1,700 passengers, actually carrying around 4,500 passengers at peak hours.[199]

The Chhatrapati Shivaji Terminus, formerly known as Victoria Terminus, is the headquarters of the Central Railway and a UNESCO World Heritage Site

Mumbai Metro is an underground and elevated rapid transit system currently under construction.[200] The Mumbai Monorail, currently under construction, will eventually run from Jacob Circle to Wadala.[201]

Mumbai is well connected to most parts of India by the Indian Railways. Long-distance trains originate from Chhatrapati Shivaji Terminus, Dadar Station, Lokmanya Tilak Terminus, Mumbai Central Station, Bandra Terminus and Andheri.[202]

Air

[203] The Chhatrapati Shivaji International Airport (formerly Sahar International Airport) is the main aviation hub in the city and the busiest airport in India in term of passenger traffic.[203] In 2007, it catered to over 25 million passengers. An upgrade plan initiated in 2006, targeted at increasing the capacity of the airport to handle up to 40 million passengers annually by 2010, is expected to be completed on time.[204]

Chhatrapati Shivaji International Airport is currently India's busiest airport in term of passenger traffic.

The proposed Navi Mumbai International Airport to be built in the Kopra-Panvel area has been sanctioned by the Indian Government and will help relieve the increasing traffic burden on the existing airport.[205]

The Juhu aerodrome was India's first airport, and now hosts a flying club and a heliport.[206]

Sea

Mumbai is served by two major ports, Mumbai Port Trust and Jawaharlal Nehru Port Trust, which lies just across the creek in Navi Mumbai.[207] Mumbai Port has one of the best natural harbours in the world, and has extensive wet and dry dock accommodation facilities.[208] Jawaharlal Nehru Port, commissioned on 26 May 1989, is the busiest and most modern major port in India.[209] It handles 55–60% of the country's total containerized cargo.[210]

Mumbai is the headquarters of the Western Naval Command, and also an important base for the Indian Navy.[104] Ferries from Ferry Wharf in Mazagaon allow access to islands near the city.[211]

Utility services

Under colonial rule, tanks were the only source of water in Mumbai. Many localities have been named after them. The BMC supplies potable water to the city from six lakes,[212] [213] most of which comes from the Tulsi and Vihar lakes.[111] The Tansa lake supplies water to the western suburbs and parts of the island city along the Western Railway.[214] The water is filtered at Bhandup,[214] which is Asia's largest water filtration plant.[215] India's first underground water tunnel is being built in Mumbai.[216] About 700 million litres of water, out of a daily supply of 3500 million litres, is lost by way of water thefts, illegal connections and leakages, per day in Mumbai.[217] Almost all of Mumbai's daily refuse of 7,800 metric tonnes, of which 40 metric tonnes is plastic waste,[218] is transported to dumping grounds in Gorai in the northwest, Mulund in the northeast, and Deonar in the east.[219] Sewage treatment is carried out at Worli and Bandra, and disposed off by two independent marine outfalls of 3.4 km (2.1 mi) and 3.7 km (2.3 mi) at Bandra and Worli respectively.[220]

BMC headquarters

Electricity is distributed by Brihanmumbai Electric Supply and Transport (BEST) in the island city, and by Reliance Energy, Tata Power, and Mahavitaran (Maharashtra State Electricity Distribution Co. Ltd) in the suburbs.[221] Consumption of electricity is growing faster than production capacity.[222] The largest telephone service provider is the state-owned MTNL, which held a monopoly over fixed line and cellular services up until 2000, and provides fixed line as well as mobile WLL services.[223] Cell phone coverage is extensive, and the main service providers are Vodafone Essar, Airtel, MTNL, BPL group, Reliance Communications, Idea Cellular and Tata Indicom. Both GSM and CDMA services are available in the city.[224] MTNL and Airtel also provide broadband internet service.[225] [226]

Demographics

Population growth		
Census	**Pop.**	**%±**
1971	5970575	—
1981	8243405	38.1%
1991	9925891	20.4%
2001	11914398	20.0%
Source:MMRDA[227] Data is based on Government of India Census.		

According to the 2001 census, the population of Mumbai was 11,914,398,[228] According to extrapolations carried out by the World Gazetteer in 2008, Mumbai has a population of 13,662,885[229] and the Mumbai Metropolitan Area has a population of 21,347,412.[230] The population density is estimated to be about 22,000 persons per square kilometre. Per 2001 census, Greater Mumbai, the area under the administration of BMC, has literacy rate of 77.45%,[231] higher than the national average of 64.8%.[232] The sex ratio was 774 (females per 1,000 males) in the island city, 826 in the suburbs, and 811 as a whole in Greater Mumbai,[231] all numbers lower than the national average of 933 females per 1,000 males.[233] The low sex ratio is due to a large number of male migrants who come to the city to work.[234] In 2008 crime rate increased by 5.4 per cent. Mumbai has registered the poorest conviction rate in the country during 2008. More frequent crimes included murder, attempt to murder, culpable homicide, dowry deaths, abduction, kidnapping, rape, arson, riots, dacoity and robbery.[235]

The religions represented in Mumbai include Hindus (67.39%), Muslims (18.56%), Buddhists (5.22%), Jains (3.99%), Christians (3.72%), Sikhs (0.58%), with Parsis and Jews making up the rest of the population.[236] The linguistic/ethnic demographics are: Maharashtrians (60%), Gujaratis (19%), with the rest hailing from other parts of India.[237] The oldest Muslim communities in Mumbai include the Dawoodi Bohras, Khojas, and Konkani Muslims.[238] Native Christians include East Indian Catholics who were converted by the Portuguese, during the 16th century.[239] The city also has a small native Bene Israeli Jewish community, who migrated from the Persian Gulf or Yemen, probably 1600 years ago.[240] Parsis migrated to India from Persia. Today, there are about 80,000 Parsis in Mumbai.[241]

Residents of Mumbai call themselves *Mumbaikar*, *Mumbaite* or *Bombayite*. Mumbai has a large polyglot population like any other metropolitan city of India. Sixteen major languages of India are spoken in Mumbai, the official language is Marathi,other spoken languages are Hindi, Gujarati and English.[242] English is extensively spoken and is the principal language of the city's white collar workforce. A colloquial form of Hindi, known as *Bambaiya*—a blend of Marathi, Hindi, Indian English and some invented words—is spoken on the streets.[243]

A street in Dharavi, one of the largest slums in the world

Mumbai suffers from the same major urbanisation problems seen in many fast growing cities in developing countries: widespread poverty and unemployment, poor public health and poor civic and educational standards for a large section of the population. With available space at a premium, Mumbai residents often reside in cramped, relatively expensive housing, usually far from workplaces, and therefore requiring long commutes on crowded mass transit, or clogged roadways. Many of them live in close proximity to bus or train stations although suburban residents spend significant time travelling southward to the main commercial district.[244] About 60% of Mumbai's population lives in slums.[245] Dharavi, Asia's second largest slum[246] is located in central Mumbai and houses 800,000 people.[247] The number of migrants to Mumbai from outside Maharashtra during the 1991–2001 decade was 1.12 million, which amounted to 54.8% of the net addition to the population of Mumbai.[248] In 2007, the crime rate (crimes booked under Indian Penal Code) in Mumbai was 186.2 per 100,000 people, which was slightly higher than the national average of 175.1, but much lower than the average crime rate of 312.3 in cities with more than one million people in the country.[249] The city's main and oldest jail is the Arthur Road Jail.[250]

Culture

Asiatic Society of Bombay is one of the oldest public libraries in the city

Mumbai's culture is a blend of traditional festivals, food, music, and theatres. The city offers a cosmopolitan and diverse lifestyle with a variety of food, entertainment and night life, available in a form and abundance comparable to that in other world capitals. Mumbai's history as a major trading centre has led to a diverse range of cultures, religions and cuisines coexisting in the city. This unique blend of cultures is due to the migration of people from all over India since the British period.[251]

Mumbai is the birthplace of Indian cinema[252] —Dadasaheb Phalke laid the foundations with silent movies followed by Marathi talkies—and the oldest film broadcast took place in the early 20th century.[253] Mumbai also has a large number of cinema halls that feature Bollywood, Marathi and Hollywood movies. The world's largest IMAX dome theatre is in the Wadala neighbourhood.[254] The Mumbai International Film Festival[255] and the award ceremony of the Filmfare Awards, the oldest and prominent film awards given for Hindi film industry in India, are held in Mumbai.[256] Despite most of the professional theatre groups that formed during the British Raj having disbanded by the 1950s, Mumbai has developed a thriving "theatre movement" tradition in Marathi, Hindi, English and other regional languages.[257] [258]

Contemporary art is featured in both government-funded art spaces and private commercial galleries. The government-funded institutions include the Jehangir Art Gallery and the National Gallery of Modern Art. Built in 1833, the Asiatic Society of Bombay is one of the oldest public libraries in the city.[259] The Chhatrapati Shivaji Maharaj Vastu Sangrahalaya (formerly The Prince of Wales Museum) is a renowned museum in South Mumbai which houses rare ancient exhibits of Indian history.[260] Mumbai has a zoo named Jijamata Udyaan (formerly Victoria Gardens), which also harbours a garden.[261] The rich literary traditions of the city have been highlighted internationally by Booker Prize winners Salman Rushdie, Aravind Adiga. Marathi literature has been modernised in the works of Mumbai based authors such as Mohan Apte, Anant Kanekar, and Gangadhar Gadgil, and is promoted through an annual Sahitya Akademi Award, a literary honour bestowed by India's National Academy of Letters.[262]

The Elephanta Caves, UNESCO World Heritage Site, feature rock-cut Shaivistic statues.

The architecture of the city is a blend of Gothic Revival, Indo-Saracenic, Art Deco, and other contemporary styles.[263] Most of the buildings during the British period, such as the Victoria Terminus and Bombay University, were built in Gothic Revival style.[264] Their architectural features include a variety of European influences such as German gables, Dutch roofs, Swiss timbering, Romance arches, Tudor casements, and traditional Indian features.[265] There are also a few Indo-Saracenic styled buildings such as the Gateway of India.[266] Art Deco styled landmarks can be found along the Marine Drive and west of the Oval Maidan.[267] Mumbai has the second largest number of Art Deco buildings in the world after Miami.[263] In the newer suburbs, modern buildings dominate the landscape. Mumbai has by far the largest number of skyscrapers in India, with 956 existing buildings and 272 under construction as of 2009.[268] The Mumbai Heritage Conservation Committee (MHCC), established in 1995, formulates special regulations and by-laws to assist in the conservation of the city's heritage structures.[263] Mumbai has two UNESCO World Heritage Sites, the Chhatrapati Shivaji Terminus and the Elephanta Caves.[269] Popular tourist attractions in the city are Nariman Point, Girgaum Chowpatti, Juhu Beach, and Marine Drive. Essel World is a theme park and amusement centre situated close to Gorai Beach,[270] and includes Asia's largest theme water park, Water Kingdom.[271]

Ganesh Chaturthi, a popular festival in Mumbai involves worship of idols of Ganesha.

Mumbai residents celebrate both Western and Indian festivals. Diwali, Holi, Eid, Christmas, Navratri, Good Friday, Dussera, Moharram, Ganesh Chaturthi, Durga Puja and Maha Shivratri are some of the popular festivals in the city. The Kala Ghoda Arts Festival is an exhibition of a world of arts that encapsulates works of artists in the fields of music, dance, theater, and films.[272] A week long annual fair known as Bandra Fair, starting on the following Sunday after September 8, is celebrated by people of all faiths, to commemorate the Nativity of Mary, mother of Jesus, on September 8.[273] The Banganga Festival is a two-day music festival, held annually in the month of January, which is organised by the Maharashtra Tourism Development Corporation (MTDC) at the historic Banganga Tank in Mumbai.[274] The Elephanta Festival—celebrated every February on the Elephanta Islands—is dedicated to classical Indian dance and music and attracts performers from across the country.[275] Public holidays specific to the city and the state include Maharashtra Day on 1 May, to celebrate the formation of Maharashtra state on 1 May 1960.[276] [277]

Sister cities

Mumbai has sister city agreements with the following cities:[145]

- London, United Kingdom
- Los Angeles, United States[278]
- Berlin, Germany
- Stuttgart, Germany[279]
- Saint Petersburg, Russia

- ● Yokohama, Japan[280] [281]

Media

Bollywood is based in Mumbai

Mumbai has numerous newspaper publications, television and radio stations. Popular English language newspapers published and sold in Mumbai include the *Times of India*, *Mid-day*, *Hindustan Times*, *DNA*, and *Indian Express*. Popular Marathi language newspapers are *Navakal*, *Maharashtra Times*, *Loksatta*, *Lokmat* and *Sakaal*. Newspapers are also printed in other Indian languages.[282] Mumbai is home to Asia's oldest newspaper, *Bombay Samachar*, which has been published in Gujarati since 1822.[283] *Bombay Durpan* the first Marathi newspaper was started by Balshastri Jambhekar in Mumbai in 1832.[284]

Numerous Indian and international television channels can be watched in Mumbai through one of the Pay TV companies or the local cable television provider. The metropolis is also the hub of many international media corporations, with many news channels and print publications having a major presence. The national television broadcaster, Doordarshan, provides two free terrestrial channels,[285] while three main cable networks serve most households.[286] The wide range of cable channels available includes ESPN, Star Sports, Zee Marathi, ETV Marathi, DD Sahyadri, Mee Marathi, Zee Talkies, Zee TV,ETV Urdu, STAR Plus and news channels such as Star Majha. News channels entirely dedicated to Mumbai include Sahara Samay Mumbai. Satellite television (DTH) has yet to gain mass acceptance, due to high installation costs.[287] Prominent DTH entertainment services in Mumbai include Dish TV and Tata Sky.[288] There are twelve radio stations in Mumbai, with nine broadcasting on the FM band, and three All India Radio stations broadcasting on the AM band.[289] Mumbai also has access to Commercial radio providers such as WorldSpace, Sirius and XM.[290] The Conditional Access System (CAS) started by the Union Government in 2006 met a poor response in Mumbai due to competition from its sister technology Direct-to-Home (DTH) transmission service.[291]

Bollywood, the Hindi film industry based in Mumbai, produces around 150–200 films every year.[292] The name Bollywood is a portmanteau of Bombay and Hollywood.[293] The 2000s saw a growth in Bollywood's popularity overseas. This led filmmaking to new heights in terms of quality, cinematography and innovative story lines as well as technical advances such as special effects and animation.[294] Studios in Goregaon, including Film City, are the location for most movie sets.[295] The Marathi film industry is also based in Mumbai.[296]

Education

Rajabai Clock Tower at the University of Mumbai

Schools in Mumbai are either "municipal schools" (run by the BMC) or private schools (run by trusts or individuals), which in some cases receive financial aid from the government.[297] The schools are affiliated either with the Maharashtra State Board (MSBSHSE), the all-India Council for the Indian School Certificate Examinations (CISCE), the Central Board for Secondary Education (CBSE) or the National Institute of Open Schooling (NIOS) boards.[298] Marathi or English is the usual language of instruction.[299] The government run public schools lack many facilities, but are the only option for poorer residents who cannot afford the more expensive private schools.[300]

Under the 10+2+3/4 plan, students complete ten years of schooling and then enroll for two years in Junior College, where they select one of three streams: arts, commerce, or science.[301] This is followed by either a general degree course in a chosen field of study, or a professional degree course, such as law, engineering and medicine.[302] Most colleges in the city are affiliated with the University of Mumbai, one of the largest universities in the world in terms of the number of graduates.[303] The Indian Institute of Technology (Bombay),[304] Veermata Jijabai Technological Institute (VJTI),[305] University Institute of Chemical Technology (UICT)[306] which are India's premier engineering and technology schools, and SNDT Women's University are the other autonomous universities in Mumbai.[307] Mumbai is also home to National Institute of Industrial Engineering (NITIE), Jamnalal Bajaj Institute of Management Studies (JBIMS), S P Jain Institute of Management and Research and several other management schools.[308] Government Law College and Sydenham College, respectively the oldest law and commerce colleges in India, are based in Mumbai.[309] [310] The Sir J. J. School of Art is Mumbai's oldest art institution.[311]

Mumbai is home to two prominent research institutions: the Tata Institute of Fundamental Research (TIFR), and the Bhabha Atomic Research Centre (BARC).[312] The BARC operates CIRUS, a 40 MW nuclear research reactor at their facility in Trombay.[313]

Sports

Brabourne Stadium, one of the oldest cricket stadiums in the country.

Cricket is the most popular sport in the city (and the country). Due to a shortage of grounds, various modified versions (generally referred to as gully cricket) are played everywhere. Mumbai is also home to the Board of Control for Cricket in India (BCCI)[314] and Indian Premier League(IPL).[315] The Mumbai cricket team represents the city in the Ranji Trophy and has won 39 titles, the most by any team.[316] The city is also represented by the Mumbai Indians in the Indian Premier League and by the Mumbai Champs in the Indian Cricket League. The city has two international cricket grounds, the Wankhede Stadium and the Brabourne Stadium.[317] The biggest cricketing event to be staged in the city so far was the 2006 ICC Champions Trophy final which was played at the Brabourne Stadium.[318] Eminent cricketers from Mumbai include Sachin Tendulkar[319] and Sunil Gavaskar.[320]

Football (soccer) is another popular sport in the city, with the FIFA World Cup and the English Premier League being followed widely.[321] In the I-League, Mumbai is represented by three teams, Mumbai FC,[322] Mahindra United[323] and Air-India.[324] Field hockey has declined in popularity, due to the rise of cricket. Mumbai is home to the Maratha Warriors, the only team from Maharashtra competing in the Premier Hockey League (PHL).[325] Every February, Mumbai holds derby races at the Mahalaxmi Racecourse. Mcdowell's Derby is also held in February at the Turf club in Mumbai.[326] Interest in Formula One racing has been rising in recent years,[327] and in 2008, the Force India F1 team car was unveiled in Mumbai.[328] In March 2004, the Mumbai Grand Prix was part of the F1 powerboat world championship.[329] In 2004, the annual Mumbai Marathon was established in a bid to bring the sports discipline to the Indian public.[330] Mumbai has also played host to the Kingfisher Airlines Tennis Open, an International Series tournament of the ATP World Tour, in 2006 and 2007.[331]

References

- Bapat, Jyotsna (2005). *Development projects and critical theory of environment*. SAGE. ISBN 9780761933571.
- Baptista, Elsie Wilhelmina (1967). *The East Indians: Catholic Community of Bombay, Salsette and Bassein*. Bombay East Indian Association.
- Bates, Crispin (2003). *Community, Empire and Migration: South Asians in Diaspora*. Orient Blackswan. ISBN 9788125024828.
- Brunn, Stanley; Williams, Jack Francis; Zeigler, Donald (2003). *Cities of the World: World Regional Urban Development* (Third ed.). Rowman & Littlefield Publishers, Inc..
- Campbell, Dennis (2008). *International Telecommunications Law [2008]*. **II**. ISBN 143571699X.
- *Census of India, 1961*. **5**. Office of the Registrar General (India). 1962.
- Carsten, F. L. (1961). *The New Cambridge Modern History (The ascendancy of France 1648-88)*. **V**. Cambridge University Press Archive. ISBN 9780521045445.
- Chaudhuri, Asha Kuthari (2005). "Introduction: Modern Indian Drama" [332]. *Mahesh Dattani: An Introduction* [333]. Contemporary Indian Writers in English. Foundation Books. ISBN 8175962607. Retrieved 2009-04-26.
- Chittar, Shantaram D. (1973). *The Port of Bombay: a brief history*. Bombay Port Trust.
- Datta, Kavita; Jones, Gareth A. (1999). *Housing and finance in developing countries*. Volume 7 of Routledge studies in development and society (illustrated ed.). Routledge. ISBN 9780415172424.
- David, M. D. (1973). *History of Bombay, 1661–1708*. University of Bombay.
- David, M. D. (1995). *Bombay, the city of dreams: a history of the first city in India*. Himalaya Publishing House.
- Davis, Mike (2006). *Planet of Slums [« Le pire des mondes possibles : de l'explosion urbaine au bidonville global »]*. Paris: La Découverte. ISBN 978-2-7071-4915-2.
- Dwivedi, Sharada; Mehrotra, Rahul (2001). *Bombay: The Cities Within*. Eminence Designs.
- *Environment and urbanization* [334]. **v. 14, no. 1**. International Institute for Environment and Development. April 2002. ISBN 9781843692232. Retrieved 2009-08-29.
- "Executive Summary on Comprehensive Transportation Study for MMR" [335] (PDF). Mumbai Metropolitan Region Development Authority (MMRDA). Retrieved 2009-08-28.
- Farooqui, Amar (2006). *Opium city: the making of early Victorian Bombay*. Three Essays Press. ISBN 9788188789320.
- Fortescue, J. W. (2008). *A History of the British Army*. **III**. Read Books. ISBN 9781443777681.
- Fuller, Christopher John; Bénéï, Véronique (2001). *The everyday state and society in modern India*. C. Hurst & Co. Publishers. ISBN 9781850654711.
- Ganti, Tejaswini (2004). "Introduction" [336]. *Bollywood: a guidebook to popular Hindi cinema*. Routledge. ISBN 0415288541.
- *Greater Bombay District Gazetteer*. Maharashtra State Gazetteers. **v. 27, no. 1**. Gazetteer Department (Government of Maharashtra). 1960.
- Ghosh, Amalananda (1990). *An Encyclopaedia of Indian Archaeology*. Brill.

- Guha, Ramachandra (2007). *India after Gandhi*. HarperCollins.
- Gupta, Om (2006). *Encyclopaedia of Journalism and Mass Communication* [337]. Kalpaz Publications.
- Hansen, Thomas Blom (2001). *Wages of violence: naming and identity in postcolonial Bombay* [338]. Princeton University Press. ISBN 9780691088402. Retrieved 2009-08-16.
- Huda, Anwar (2004). *The Art and Science of Cinema* [339]. Atlantic Publishers & Distributors. ISBN 9788126903481. Retrieved 2008-06-11.
- Jha, Subhash K. (2005). *The Essential Guide to Bollywood*. Roli Books. ISBN 8174363785.
- Keillor, Bruce David (2007). *Marketing in the 21st Century: New world marketing*. **1**. Praeger. ISBN 0275992764.
- Kelsey, Jane (2008). *Serving Whose Interests?: The Political Economy of Trade in Services Agreements*. Taylor & Francis. ISBN 9780415448215.
- Khalidi, Omar (2006). *Muslims in the Deccan: a historical survey*. Global Media Publications. ISBN 9788188869138.
- Kothari, Rajni (1970). *Politics in India*. Orient Longman.
- Krishnamoorthy, Bala (2008). *Environmental Management: Text And Cases*. PHI Learning Pvt. Ltd.. ISBN 9788120333291.
- Kumari, Asha (1990). *Hinduism and Buddhism*. Vishwavidyalaya Prakashan. ISBN 8171240607.
- *Lok Sabha debates*. New Delhi: Lok Sabha Secretariat. 1998.
- Machado, José Pedro (1984). "Bombaim" (in Portuguese). *Dicionário Onomástico Etimológico da Língua Portuguesa*. **I**. *Editorial Confluência*.
- Mehta, Suketu (2004). *Maximum City: Bombay Lost and Found*. Alfred A Knopf. ISBN 0-375-40372-8.
- *Metropolitan planning and management in the developing world: spatial decentralization policy in Bombay and Cairo*. United Nations Centre for Human Settlements. 1993. ISBN 9789211312331.
- Misra, Satish Chandra (1982). *The Rise of Muslim Power in Gujarat: A History of Gujarat from 1298 to 1442*. Munshiram Manoharlal Publishers.
- Morris, Jan; Winchester, Simon (2005) [1983]. *Stones of empire: the buildings of the Raj* (reissue, illustrated ed.). Oxford University Press. ISBN 9780192805966.
- "Mumbai Plan" [340]. Department of Relief and Rehabilitation (Government of Maharashtra). Retrieved 2009-04-29.
- Naravane, M. S. (2007). *Battles of the honourable East India Company: making of the Raj*. APH Publishing. ISBN 9788131300343.
- O'Brien, Derek (2003). *The Mumbai Factfile*. Penguin Books. ISBN 9780143029472.
- "Office of the Commissioner of Police, Mumbai" [341] (PDF, 1.18 MB). Mumbai Police. Retrieved 2009-06-15.
- Patel, Sujata; Masselos, Jim, eds (2003). "Bombay and Mumbai: Identities, Politics and Populism". *Bombay and Mumbai. The City in Transition*. Delhi, India: The Oxford University Press. ISBN 0195677110.
- Pai, Pushpa (2005). "Multilingualism, Multiculturalism and Education: Case Study of Mumbai City" [342]. in Cohen, James; McAlister, Kara T.; Rolstad, Kellie; MacSwan, Jeff (PDF). *Proceedings of the 4th International Symposium on Bilingualism*. Cascadilla Press. pp. 1794–1806. Retrieved 2009-09-05.
- Patil, R.P. (1957). *The mangroves in Salsette Island near Bombay*. Calcutta: Proceedings of the Symposium on Mangrove Forest.
- Phadnis, Aditi. *Business Standard Political Profiles: Of Cabals and Kings*. Business Standard.
- "Population and Employement profile of Mumbai Metropolitan Region" [343] (PDF). Mumbai Metropolitan Region Development Authority (MMRDA). Retrieved 2009-05-03.
- *Proceedings of the Indian National Science Academy*. **65**. Indian National Science Academy. 1999.
- Rana, Mahendra Singh (2006). *India votes: Lok Sabha & Vidhan Sabha elections 2001-2005*. Sarup & Sons. ISBN 9788176256476.

- Rohli, Robert V.; Vega, Anthony J. (2007). *Climatology* (illustrated ed.). Jones & Bartlett Publishers. ISBN 9780763738280.
- Saini, A.K.; Chand; Hukam. *History Of Midieval India*. Anmol Publications. ISBN 9788126123131.
- Singh, K. S.; B. V. Bhanu, B. R. Bhatnagar, Anthropological Survey of India, D. K. Bose, V. S. Kulkarni, J. Sreenath (2004). *Maharashtra*. **XXX**. Popular Prakashan. ISBN 9788179911020.
- Shirodkar, Prakashchandra P. (1998). *Researches in Indo-Portuguese history*. **2**. Publication Scheme. ISBN 9788186782156.
- Swaminathan, R.; Goyal, Jaya (2006). *Mumbai vision 2015: agenda for urban renewal*. Macmillan India in association with Observer Research Foundation.
- Strizower, Schifra. (1971). *The children of Israel: the Bene Israel of Bombay*. B. Blackwell.
- *The Gazetteer of Bombay City and Island*. Gazetteers of the Bombay Presidency. **2**. Gazetteer Department (Government of Maharashtra). 1978.
- "The Mumbai Municipal Corporation Act, 1888" [344] (PDF). State Election Commissioner (Government of Maharashtra). Retrieved 2009-05-03.
- Vilanilam, John V. (2005). *Mass communication in India: a sociological perspective* (illustrated ed.). SAGE. ISBN 9780761933724.
- Wasko, Janet (2003). *How Hollywood works*. SAGE. ISBN 0761968148.
- *WMO bulletin*. **49**. World Meteorological Organization. 2000.
- Yimene, Ababu Minda (2004). *An African Indian Community in Hyderabad: Siddi Identity, Its Maintenance and Change*. Cuvillier Verlag. ISBN 3865372066.
- Yule, Henry; Burnell (1996) [1939]. *A glossary of colloquial Anglo-Indian words and phrases: Hobson-Jobson* (2 ed.). Routledge. ISBN 9780700703210.
- Zakakria, Rafiq; Indian National Congress (1985). *100 glorious years: Indian National Congress, 1885–1985*. Reception Committee, Congress Centenary Session.

Further reading

- Agarwal, Jagdish (1998). *Bombay — Mumbai: A Picture Book*. Wilco Publishing House. ISBN 81-87288-35-3.
- Chaudhari, K.K (1987). *History of Bombay*. Modern Period Gazetteers Department (Government of Maharashtra).
- Contractor, Behram (1998). *From Bombay to Mumbai*. Oriana Books.
- Cox, Edmund Charles (1887) (PDF, 32 MB). *A short history of the Bombay Presidency* [345]. Thacker and Company. Retrieved 2009-06-08.
- Hunter, William Wilson; Cotton, James Sutherland; Burn, Richard; Meyer, William Stevenson; Great Britain India Office (1909). "Berhampore — Bombay" [346]. *The Imperial Gazetteer of India* [347]. **8**. Oxford: Clarendon Press. Retrieved 2009-06-08.
- Katiyar, Arun; Bhojani, Namas (1996). *Bombay, A Contemporary Account*. HarperCollins. ISBN 81-7223-216-0.
- MacLean, James Mackenzie (1876). *A Guide to Bombay: Historical, Statistical, and Descriptive*. Bombay Gazette steam Press.
- Mappls (1999). *Satellite based comprehensive maps of Mumbai*. CE Info Systems Limited. ISBN 81-901108-0-2.
- *Our Greater Bombay*. Maharashtra State Bureau of Textbook Production and Curriculum Research. 1990.
- Patel, Sujata; Thorner, Alice (1995). *Bombay, Metaphor for Modern India*. Oxford University Press. ISBN 0-19-563688-0.
- Tindall, Gillian (1992). *City of Gold*. Penguin Books. ISBN 0-14-009500-4.
- Virani, Pinki (1999). *Once was Bombay*. Viking. ISBN 0-670-88869-9.

External links

- Official site of the Municipal Corporation of Greater Mumbai [348]
- Official City Report [340]
- Mumbai travel guide from Wikitravel

References

[1] "World: largest cities and towns and statistics of their population (2010)" (http://world-gazetteer.com/wg.php?x=&men=gcis&lng=en&des=wg&srt=npan&col=abcdefghinoq&msz=1500&pt=c&va=&srt=pnan). World Gazetteer. . Retrieved 2009-04-28.

[2] "India: metropolitan areas" (http://www.world-gazetteer.com/wg.php?x=&men=gcis&lng=en&dat=80&geo=-104&srt=pnan&col=aohdq&msz=1500&va=&pt=a). World Gazetteer. . Retrieved 2010-03-10.

[3] http://www.mcgm.gov.in

[4] "Population of urban agglomerations with 750,000 inhabitants or more in 2007 (thousands) 1950-2025 (India)" (http://esa.un.org/unup/index.asp?panel=2). Department of Economic and Social Affairs (United Nations). . Retrieved 2009-06-09.

[5] "GAWC World Cities Ranking List" (http://www.diserio.com/gawc-world-cities.html). Diserio.com. . Retrieved 2010-05-05.

[6] "India needs cities network for easy rural-urban shift - Economy and Politics" (http://www.livemint.com/2009/08/03224002/India-needs-cities-network-for.html). livemint.com. 2009-08-03. . Retrieved 2010-05-05.

[7] Dwivedi & Mehrotra 2001, p. 28

[8] "Bombay: History of a City" (http://www.bl.uk/learning/histcitizen/trading/bombay/history.html). British Library. . Retrieved 2008-11-08.

[9] "Mumbai Urban Infrastructure Project" (http://mmrdamumbai.org/projects_muip.htm). Mumbai Metropolitan Region Development Authority (MMRDA). . Retrieved 2008-07-18.

[10] "Navi Mumbai International Airport" (http://img214.imageshack.us/img214/2299/dscn7619ql4.jpg) (JPG). City and Industrial Development Corporation (CIDCO). . Retrieved 2008-07-18.

[11] Chittar 1973, p. 6

[12] Yule & Burnell 1996, p. 103 (http://books.google.co.in/books?id=20pdFRekGvMC&pg=PA103&dq=mombayn++Bombay+Portuguese&lr=#v=onepage&q=mombayn Bombay Portuguese&f=false)

[13] Greater Bombay District Gazetteer 1960, p. 6

[14] Patel & Masselos 2003, p. 4

[15] Mehta 2004, p. 130

[16] Hansen 2001, p. 1

[17] "Mumbai (Bombay) and Maharashtra" (http://www.fodors.com/world/asia/india/mumbai-bombay-and-maharashtra/more.html). Fodor's. . Retrieved 2009-08-24.

[18] Shirodkar 1998, p. 7

[19] Machado 1984, pp. 265–266

[20] Shirodkar 1998, p. 3

[21] Shirodkar 1998, pp. 4–5

[22] Yule & Burnell 1996, p. 102 (http://books.google.co.in/books?id=20pdFRekGvMC&pg=PA102&dq=mombayn++Bombay+Portuguese&lr=#v=onepage&q=mombayn Bombay Portuguese&f=false)

[23] Shirodkar 1998, p. 2

[24] Yule & Burnell 1996, p. 104 (http://books.google.co.in/books?id=20pdFRekGvMC&pg=PA104&dq=mombayn++Bombay+Portuguese&lr=#v=onepage&q=mombayn Bombay Portuguese&f=false)

[25] Farooqui 2006, p. 1

[26] Ghosh 1990, p. 25

[27] Greater Bombay District Gazetteer 1960, p. 5

[28] David 1995, p. 5

[29] "Kanheri Caves" (http://asi.nic.in/asi_monu_tktd_maha_kanhericaves.asp). Archaeological Survey of India (ASI). . Retrieved 2008-10-17.

[30] Kumari 1990, p. 37

[31] David 1973, p. 8

[32] Greater Bombay District Gazetteer 1960, pp. 127–150

[33] Dwivedi & Mehrotra 2001, p. 79

[34] "The Slum and the Sacred Cave" (http://www.ldeo.columbia.edu/edu/eesj/gradpubs/GeneralMags/Patel_Archaeology_SlumandSacredCave_0607.pdf) (PDF). Lamont-Doherty Earth Observatory (Columbia University). p. 5. . Retrieved 2008-10-12.

[35] "World Heritage Sites — Elephanta Caves" (http://asi.nic.in/asi_monu_whs_elephanta.asp). Archaeological Survey of India. . Retrieved 2008-10-22.

[36] Dwivedi, Sharada (2007-09-26). "The Legends of Walkeshwar" (http://cities.expressindia.com/fullstory.php?newsid=101117). *Mumbai Newsline*. Express Group. . Retrieved 2009-01-31.
[37] Agarwal, Lekha (2007-06-02). "What about Gateway of India, Banganga Tank?" (http://cities.expressindia.com/fullstory.php?newsid=239318). *Mumbai Newsline*. Express Group. . Retrieved 2009-01-31.
[38] Dwivedi & Mehrotra 2001, p. 51
[39] Maharashtra 2004, p. 1703
[40] David 1973, p. 14
[41] David 1995, p. 12
[42] Khalidi 2006, p. 24
[43] Misra 1982, p. 193
[44] Misra 1982, p. 222
[45] David 1973, p. 16
[46] "Mughal Empire" (http://www.sscnet.ucla.edu/southasia/History/Mughals/mughals.html). Department of Social Sciences (University of California). . Retrieved 2009-05-22.
[47] Greater Bombay District Gazetteer 1960, p. 166
[48] Greater Bombay District Gazetteer 1960, p. 169
[49] David 1995, p. 19
[50] Shukla, Ashutosh (2008-05-12). "Relishing a Sunday feast, but only once in a year" (http://www.dnaindia.com/mumbai/report_relishing-a-sunday-feast-but-only-once-in-a-year_1163869). *Daily News and Analysis* (DNA). . Retrieved 2009-09-02.
[51] D'Mello, Ashley (2008-06-09). "New life for old church records" (http://timesofindia.indiatimes.com/Cities/Mumbai/New_life_for_old_church_records/articleshow/3112498.cms). The Times of India. . Retrieved 2009-09-02.
[52] "Glorious past" (http://www.expressindia.com/latest-news/glorious-past/233152/). Express India. 2008-10-28. . Retrieved 2009-06-17.
[53] "Catherine of Bragança (1638 - 1705)" (http://www.bbc.co.uk/dna/h2g2/A2998461). BBC. . Retrieved 2008-11-05.
[54] The Gazetteer of Bombay City and Island 1978, p. 54
[55] Dwivedi & Mehrotra 2001, p. 20
[56] David 1973, p. 410
[57] Yimene 2004, p. 94
[58] Ganley, Colin C. (2007) (PDF, 113 KB). *Security, the central component of an early modern institutional matrix; 17th century Bombay's Economic Growth* (http://www.isnie.org/assets/files/papers2007/ganley.pdf). International Society for New Institutional Economics (ISNIE). p. 13. . Retrieved 2008-11-06.
[59] Carsten 1961, p. 427
[60] David 1973, p. 179
[61] Nandgaonkar, Satish (2003-03-22). "Mazgaon fort was blown to pieces – 313 years ago" (http://cities.expressindia.com/fullstory.php?newsid=47106). *Indian Express* (Express Group). . Retrieved 2008-09-20.
[62] History of Midieval India, p. 126
[63] Dwivedi & Mehrotra 2001, p. 32
[64] Fortescue 2008, p. 145
[65] Naravane 2007, p. 56
[66] Naravane 2007, p. 63
[67] Naravane 2007, pp. 80–82
[68] Greater Bombay District Gazetteer 1960, p. 233
[69] "Maharashtra — trivia" (http://www.maharashtratourism.gov.in/MTDC/HTML/MaharashtraTourism/Default.aspx?strpage=../MaharashtraTourism/Trivia.html). Maharashtra Tourism Development Corporation. . Retrieved 2007-12-07.
[70] Dwivedi & Mehrotra 2001, p. 127
[71] Dwivedi & Mehrotra 2001, p. 343
[72] Dwivedi & Mehrotra 2001, p. 88
[73] Dwivedi & Mehrotra 2001, p. 74
[74] "Rat Trap" (http://www.timeoutmumbai.net/aroundtown/aroundtown_preview_details.asp?code=45). *Time out (Mumbai)* (Time Out) (6). 2008-11-14. . Retrieved 2008-11-19.
[75] Dwivedi & Mehrotra 2001, p. 345
[76] Dwivedi & Mehrotra 2001, p. 293
[77] Census of India 1961, p. 23
[78] "Administration" (http://mumbaisuburban.gov.in/html/administrative_setup.htm). Mumbai Suburban District. . Retrieved 2008-11-06.
[79] Guha, Ramachandra (2003-04-13). "The battle for Bombay" (http://www.hinduonnet.com/thehindu/mag/2003/04/13/stories/2003041300240300.htm). *The Hindu*. . Retrieved 2008-11-12.
[80] Guha 2007, pp. 197–8
[81] "Samyukta Maharashtra" (http://www.maharashtra.gov.in/english/community/community_samyuktaShow.php). Government of Maharashtra. . Retrieved 2008-11-12.

[82] "Sons of soil: born, reborn" (http://www.indianexpress.com/news/sons-of-soil-born-reborn/269628/). *Indian Express Newspapers (Mumbai)*. 2008-02-06. . Retrieved on 2008-11-12.

[83] "Gujarat" (http://india.gov.in/knowindia/st_gujurat.php). Government of India. . Retrieved 2008-01-16.

[84] "Maharashtra" (http://india.gov.in/knowindia/st_maharashtra.php). Government of India. . Retrieved 2008-01-16.

[85] Desai, Geeta (2008-05-13). "BMC will give jobs to kin of Samyukta Maharashtra martyrs" (http://epaper.timesofindia.com/Repository/ml.asp?Ref=TU1JUi8yMDA4LzA1LzEzI0FyMDA1MDA=&Mode=HTML&Locale=english-skin-custom). Mumbai Mirror. . Retrieved 2008-11-16.

[86] Dwivedi & Mehrotra 2001, p. 306

[87] "About Mumbai Metropolitan Region Development Authority (MMRDA)" (http://www.mmrdamumbai.org/index.htm). Mumbai Metropolitan Region Development Authority. . Retrieved 2008-11-13.

[88] "About Navi Mumbai (History)" (http://www.nmmconline.com/english/aboutUs/about_history_Show.php). Navi Mumbai Municipal Corporation (NMMC). . Retrieved 2008-11-13.

[89] "The Great Mumbai Textile Strike... 25 Years On" (http://www.rediff.com/news/2007/jan/18sld2.htm). Rediff.com India Limited. 2007-01-18. . Retrieved 2008-11-20.

[90] "Profile of Jawaharlal Nehru Custom House (Nhava Sheva)" (http://www.jawaharcustoms.gov.in/jnch/others/profile.htm). Jawaharlal Nehru Custom House. . Retrieved 2008-11-13.

[91] "Profile" (http://mumbaisuburban.gov.in/html/profile.htm). Mumbai Suburban District. .

[92] Kaur, Naunidhi (July 05–18, 2003). "Mumbai: A decade after riots" (http://www.hinduonnet.com/fline/fl2014/stories/20030718002704100.htm). *Frontline* **20** (14). . Retrieved 2008-11-13.

[93] "1993: Bombay hit by devastating bombs" (http://news.bbc.co.uk/onthisday/hi/dates/stories/march/12/newsid_4272000/4272943.stm). *BBC News*. 1993-03-12. . Retrieved 2008-11-12.

[94] "Special Report: Mumbai Train Attacks" (http://news.bbc.co.uk/2/hi/in_depth/south_asia/2006/mumbai_train_attacks/default.stm). *BBC News*. 2006-09-30. . Retrieved 2008-08-13.

[95] Press Information Bureau (Government of India) (2008-12-11). "HM announces measures to enhance security" (http://pib.nic.in/release/release.asp?relid=45446). Press release. . Retrieved 2008-12-14.

[96] Thomas, T. (2007-04-27). "Mumbai a global financial centre? Of course!" (http://www.rediff.com/money/2007/apr/27mumbai.htm). New Delhi: Rediff. . Retrieved 2009-05-31.

[97] Shaw, Annapurna (1999). "Emerging Patterns of Urban Growth in India" (http://www.jstor.org/stable/4407880). *Economic and Political Weekly* **34** (16/17): 969–978. doi:10.2307/4407880 (inactive 2009-09-20). . Retrieved 2009-07-08.

[98] Brunn, Williams & Zeigler 2003, pp. 353–354

[99] "Mumbai Suburban" (http://www.maharashtra.nic.in/htmldocs/Activity/mumbai_sub.pdf) (PDF). National Informatics Centre (Mahrashtra State Centre). . Retrieved 2009-15-31.

[100] "MMRDA Projects" (http://www.mmrdamumbai.org/projects_muip.htm). Mumbai Metropolitan Region Development Authority (MMRDA). . Retrieved 2007-12-06.

[101] "Area and Density – Metropolitan Cities" (http://webcitation.org/5gOCPonTE) (PDF, 111 KB). Ministry of Urban Development (Government of India). p. 33. Archived from the original (http://urbanindia.nic.in/moud/theministry/subordinateoff/tcpo/AREA_POP/CHAPTER-4.PDF) on 2009-04-29. . Retrieved 2008-04-28.

[102] Mumbai Plan, 1.2 Area and Divisions

[103] Greater Bombay District Gazetteer 1960, p. 2

[104] Mumbai Plan, 1.1 Location

[105] Krishnamoorthy 2008, p. 218

[106] "Mumbai, India" (http://www.weatherbase.com/weather/weather.php3?s=030034&refer=). Weatherbase. . Retrieved 2008-03-19.

[107] Mumbai Plan, 1.3.2.2 Salsette Island

[108] Srinivasu, T.; Pardeshi, Satish. "Floristic Survey of Institute of Science, Mumbai, Maharashta State" (http://iscmumbai.maharashtra.gov.in/floristic survey.html). Government of Maharashtra. . Retrieved 2009-08-26.

[109] Bapat 2005, pp. 111–112

[110] "Municipal Corporation of Greater Mumbai Water Sector Initiatives" (http://darpg.nic.in/arpg-website/Conference/Pune/water supply initiatives.ppt) (PPT). Department of Administrative Reforms and Public Grievances (Government of India). p. 6. . Retrieved 2008-04-30.

[111] Bavadam, Lyla (February 15–28, 2003). "Encroaching on a lifeline" (http://www.hinduonnet.com/fline/fl2004/stories/20030228002609200.htm). *Frontline* (The Hindu). . Retrieved 2008-04-28.

[112] "Salient Features of Powai Lake" (http://envis.maharashtra.gov.in/envis_data/pps/pawai2.ppt) (PPT, 1.6 MB). Department of Environment (Government of Maharashtra). pp. 1–3. . Retrieved 2009-04-29.

[113] Mumbai Plan, 1.7 Water Supply and Sanitation

[114] Sen, Somit (2008-12-13). "Security web for city coastline" (http://timesofindia.indiatimes.com/Cities/Security_web_for_city_coastline/articleshow/3830390.cms). The Times of India. . Retrieved 2009-04-30.

[115] Patil 1957, pp. 45–49

[116] Mumbai Plan, 1.3.1 Soil

[117] Mumbai Plan, 1.3.2 Geology and Geomorphology

[118] Kanth, S. T. G. Raghu; Iyenagar, R. N. (2006-12-10). "Seismic Hazard estimation for Mumbai City" (http://www.scribd.com/doc/10026629/Earthquake-Hazard-Computation-for-Mumbai-Bombay-City). *Current Science* (Current Science Association) **91** (11): 1486. . Retrieved 2009-09-03. "This is used to compute the probability of ground motion that can be induced by each of the **twenty-three** known faults that exist around the city.".

[119] India Meteorological Department. *Seismic Zoning Map* (http://www.imd.ernet.in/section/seismo/static/seismo-zone.htm) [map]. Retrieved on 2008-07-20.

[120] "The Seismic Environment of Mumbai" (http://theory.tifr.res.in/bombay/physical/fault.html). Department of Theoretical Physics (Tata Institute of Fundamental Research). . Retrieved 2007-12-06.

[121] Proceedings of the Indian National Science Academy 1999, p. 210

[122] Greater Bombay District Gazetteer 1960, p. 84

[123] Mumbai Plan, 1.4 Climate and Rainfall

[124] Kishwar, Madhu Purnima (2006-07-03). "Three drown as heavy rain lashes Mumbai for the 3rd day" (http://www.dnaindia.com/report.asp?NewsID=1039257). Mumbai: *Daily News and Analysis* (DNA). . Retrieved 2009-06-15.

[125] Rohli & Vega 2007, p. 267

[126] WMO bulletin 2000, p. 346, "Bombay recorded a maximum temperature of 40.2°C on 28 March 1982, the highest since 1955."

[127] "Mumbai still cold at 8.6 degree C" (http://timesofindia.indiatimes.com/articleshow/2770007.cms). The Times of India. 2008-02-09. . Retrieved 2009-04-26.

[128] "Historical Weather for Mumbai, India" (http://www.wunderground.com/NORMS/DisplayIntlNORMS.asp?CityCode=42182&Units=both). Weather Underground. . Retrieved 27 November 2008.

[129] Swaminathan & Goyal 2006, p. 51

[130] "GDP growth: Surat fastest, Mumbai largest" (http://www.financialexpress.com/news/gdp-growth-surat-fastest-mumbai-largest/266636/). The Financial Express. 2008-01-29. . Retrieved 2009-09-05.

[131] "Fortune Global 500" (http://money.cnn.com/magazines/fortune/global500/2008/countries/India.html). *Fortune*. CNN. 2008-07-21. . Retrieved 2009-04-28.

[132] "Welcome To World Trade Centre, Mumbai" (http://www.wtcmumbai.org/). WTC Mumbai. . Retrieved 2008-02-14.

[133] Swaminathan & Goyal 2006, p. 52

[134] "The World According to GaWC 2008" (http://www.lboro.ac.uk/gawc/world2008t.html). *Globalization and World Cities Study Group and Network (GaWC)*. Loughborough University. . Retrieved 2009-05-07.

[135] http://business.rediff.com/slide-show/2010/may/07/slide-show-1-mumbai-among-top-10-expensive-office-markets.htm#contentTop

[136] Keillor 2007, p. 83

[137] "Indian Ports Association, Operational Details" (http://www.ipa.nic.in/oper.htm). Indian Ports Association. . Retrieved 2009-04-16.

[138] McDougall, Dan (2007-03-04). "Waste not, want not in the £700m slum" (http://www.guardian.co.uk/environment/2007/mar/04/india.recycling). *The Observer* (Guardian News and Media Limited). . Retrieved 2009-04-29.

[139] Wasko 2003, p. 185

[140] Jha 2005, p. 1970

[141] Kelsey 2008, p. 208

[142] "Worldwide Centres of Commerce Index 2008" (http://www.mastercard.com/us/company/en/insights/pdfs/2008/MCWW_WCoC-Report_2008.pdf) (PDF). MasterCard. p. 21. . Retrieved 2009-04-28.

[143] "In Pictures: The Top 10 Cities For Billionaires" (http://www.forbes.com/2008/04/30/billionaires-london-moscow-biz-billies-cz_cv_0430billiecities_slide_5.html?thisSpeed=15000). Forbes. . Retrieved 2009-04-28.

[144] Vorasarun, Chaniga (2008-04-30). "Cities Of The Billionaires" (http://www.forbes.com/2008/04/30/billionaires-london-moscow-biz-billies-cz_cv_0430billiecities.html). Forbes. . Retrieved 2009-04-28.

[145] "Official Website of Municipal Corporation of Greater Mumbai" (http://www.mcgm.gov.in/). Municipal Corporation of Greater Mumbai. . Retrieved 2008-07-18.

[146] "Commissioner System" (http://www.citymayors.com/government/india_government.html). .

[147] Greater Bombay District Gazetteer 1960, General Administration (Introduction) (http://www.maharashtra.gov.in/english/gazetteer/greater_bombay/generaladmin.html#1)

[148] "Collector" (http://www.maharashtra.gov.in/english/gazetteer/greater_bombay/generaladmin.html#6). Maharashtra.gov.in. . Retrieved 2010-05-05.

[149] Office of the Commissioner of Police, Mumbai, p. 2

[150] Office of the Commissioner of Police, Mumbai, pp. 7–8

[151] "About Bombay High Court" (http://bombayhighcourt.nic.in/). Bombay High Court. . Retrieved 2008-01-27.

[152] Greater Bombay District Gazetteer 1960, Judiciary (http://www.maharashtra.gov.in/english/gazetteer/greater_bombay/law.html#5)

[153] Fuller & Bénéï 2001, p. 47

[154] 100 glorious years: Indian National Congress, 1885–1985, p. 4, "The centenary of the Indian National Congress, which is being celebrated at its birthplace Bombay, is a unique event."

[155] "Congress foundation day celebrated" (http://www.hindu.com/2006/12/29/stories/2006122906471500.htm). The Hindu. 2006-12-29. . Retrieved 2008-11-12.

[156] David 1995, p. 215

[157] "Sena fate: From roar to meow" (http://timesofindia.indiatimes.com/articleshow/msid-1311115,prtpage-1.cms). The Times of India. 2005-11-29. . Retrieved 2008-11-12.

[158] "Maharashtra government 'soft' on Raj Thackeray's outfit, says BJP" (http://www.thehindu.com/2008/02/13/stories/2008021354841200.htm). The Hindu. 2008-02-13. . Retrieved 2008-04-04.

[159] Phadnis, pp. 86–87

[160] Rana 2006, pp. 315–316

[161] "Stage Set for Third Phase Polls in Maharashtra" (http://news.outlookindia.com/item.aspx?659150). *Outlook*. 2009-04-29. . Retrieved 2009-07-06.

[162] "List Of Parliamentary Constituencies" (http://eci.nic.in/eci_main/miscellaneous_statistics/ListofPC.pdf) (PDF). Election Commission of India. p. 7. . Retrieved 2009-09-04.

[163] "Congress wins five seats in Mumbai, NCP wins the sixth seat" (http://www.mumbaimirror.com/index.aspx?Page=article§name=News - City§id=2&contentid=20090516200905161552135572 0c4812d). Mumbai Mirror. 2009-05-16. . Retrieved 2009-07-06.

[164] "List of ACs and PCs" (http://ceo.maharashtra.gov.in/acs.php). Chief Electoral Officer (Government of Maharashtra). . Retrieved 2009-09-04.

[165] "Maharashtra Assemby Election 2009" (http://220.225.73.214/pdff/results.pdf). . Retrieved 18 March 2010.

[166] "Maharashtra 2004 poll outcome" (http://www.rediff.com/election/2004/oct/16kbkmaha.htm). Rediff. 2004-10-16. . Retrieved 2009-09-04.

[167] The Mumbai Municipal Corporation Act, 1888, p. 6

[168] "Corporation" (http://www.mcgm.gov.in/irj/portal/anonymous/qlcorporation). Brihanmumbai Municipal Corporation (BMC). . Retrieved 2009-06-15.

[169] "Mayor - the First Citizen of Mumbai" (http://www.mcgm.gov.in/irj/portal/anonymous/qlmayoffice). Brihanmumbai Municipal Corporation (BMC). . Retrieved 2009-05-12. "As the presiding authority at the Corporation Meetings, his/her role is confined to the four corners of the Corporation Hall. The decorative role, however, extends far beyond the city and the country to other parts of world"

[170] The Mumbai Municipal Corporation Act, 1888, p. 3

[171] "Sena's hat-trick in BMC; Congress suffers setback" (http://www.rediff.com/news/2007/feb/02poll.htm). Rediff. 2007-02-03. . Retrieved 2009-09-04.

[172] The Mumbai Municipal Corporation Act, 1888, p. 27

[173] "Development of Bus Rapid Transit System (BRTS) in Mumbai" (http://www.mmrdamumbai.org/docs/BRTS Note for web Page.doc) (DOC). Mumbai Metropolitan Region Development Authority (MMRDA). . Retrieved 2009-08-28.

[174] Ghose, Anindita (2005-08-24). "What's Mumbai without the black beetles?" (http://www.dnaindia.com/speakup/report_what-s-mumbai-without-the-black-beetles_422). Daily News and Analysis (DNA). . Retrieved 2009-08-29. "In Mumbai autos run only in the suburbs up to Mahim creek. This is probably the perfect arrangement because it is not economically viable for autos and taxis to solicit the same passengers. So autos monopolise the suburbs while taxis rule South Mumbai."

[175] "Taxi, auto fares may dip due to CNG usage" (http://timesofindia.indiatimes.com/articleshow/631726.cms). The Times of India. 2004-04-22. . Retrieved 2009-08-29.

[176] Vaswani, Karishma (2008-04-07). "Mumbai attempts 'no honking' day" (http://news.bbc.co.uk/2/hi/south_asia/7334628.stm). *BBC News* (BBC). . Retrieved 2009-08-29.

[177] Executive Summary on Comprehensive Transportation Study for MMR, p. 2–9

[178] "NH wise Details of NH in respect of Stretches entrusted to NHAI" (http://www.nhai.org/Doc/project-offer/Highways.pdf) (PDF, 62.2 KB). National Highways Authority of India (NHAI). . Retrieved 2008-07-04.

[179] Dalal, Sucheta (2000-04-01). "India's first international-class expressway is just a month away" (http://www.indianexpress.com/ie/daily/20000401/ina01059.html). The Indian Express. . Retrieved 2009-06-14.

[180] Kumar, K.P. Narayana; Chandran, Rahul (2008-03-06). "NHAI starts work on Rs 6,672 cr expressway" (http://www.livemint.com/2008/03/06231146/NHAI-starts-work-on-Rs6672-cr.html). *Mint*. . Retrieved 2009-06-14.

[181] "MSRDC - Project - Bandra Worli Sea Link" (http://www.msrdc.org/projects/bandra_worli.aspx). Maharashtra State Road Development Corporation (MSRDC). . Retrieved 2009-07-02.

[182] Mumbai Plan, 1.10 Transport and Communication Network

[183] "Organisational Setup" (http://bestundertaking.com/trans_func.asp). Brihanmumbai Electric Supply and Transport (BEST). . Retrieved 2009-06-14.

[184] Metropolitan planning and management in the developing world 1993, p. 49 (http://books.google.co.in/books?id=SD4I3CEtDz0C&pg=PA49&dq=public+transport+mumbai++Mumbai+Suburban+Railway+best+taxi&lr=#v=onepage&q=&f=false)

[185] "Composition of Bus Fleet" (http://www.bestundertaking.com/trans_engg.asp). Brihanmumbai Electric Supply and Transport (BEST). . Retrieved 2006-10-12.

[186] "Mumbai bus network tops 1000, gets new look" (http://www.screens.tv/article/11738/Mumbai_bus_network_tops_1000,_gets_new_look.html). Screens.tv. 2009-06-09. . Retrieved 2010-05-05.

[187] "Bus Transport Profile" (http://www.bestundertaking.com/trans_botright.asp). Brihanmumbai Electric Supply and Transport (BEST). . Retrieved 2009-08-28.

[188] Tembhekar, Chittaranjan (2008-08-04). "MSRTC to make long distance travel easier" (http://timesofindia.indiatimes.com/articleshow/3322572.cms). The Times of India. . Retrieved 2009-06-14.

[189] "MSRTC adds Volvo luxury to Mumbai trip" (http://timesofindia.indiatimes.com/articleshow/32792301.cms). The Times of India. 2002-12-29. . Retrieved 2009-06-14.

[190] "NMMT Volvo bus route 123 extended till Borivali from Jan 1" (http://www.mumbaipluses.com/newbombayplus/index.aspx?page=article§id=1&contentid=20100103201001041539585 74e148ffd8§xslt=&comments=true). Mumbaipluses.com. . Retrieved 2010-05-05.

[191] Seth, Urvashi (2009-03-31). "Traffic claims Mumbai darshan hot spots" (http://www.mid-day.com/news/2009/mar/310309-Mumbai-News-Mumbai-Darshan-popular-tourist-spots-traffic-congestion-Tourist.htm). MiD DAY. . Retrieved 2009-06-14.

[192] "Bus Routes Under Bus Rapid Transit System" (http://www.bestundertaking.com/TravelAsYouLike-Ticket.pdf) (PDF). Brihanmumbai Electric Supply and Transport (BEST). p. 5. . Retrieved 2009-03-23.

[193] Khanna, Gaurav. "7 Questions You Wanted to Ask About the Mumbai Metro" (http://www.businessworld.in/index.php/7-Questions-You-Wanted-to-Ask.html). Businessworld. . Retrieved 2009-08-28. "Road congestion has worsened, though 88 per cent of journeys are made by public transport."

[194] Executive Summary on Comprehensive Transportation Study for MMR, p. 2-1: "The 137% increase in cars, a 306% increase in two wheelers, the 420% increase in autos and 128% increase in taxis during 1991-2005 has created a lethal dose of traffic congestion which has categorised Mumbai as one of the congested cities in the world."

[195] "MMRDA - Projects - Skywalk" (http://www.mmrdamumbai.org/skywalk.htm). Mmrdamumbai.org. . Retrieved 2010-05-05.

[196] Executive Summary on Comprehensive Transportation Study for MMR, p. 2–14

[197] Press Information Bureau (Government of India). "Making Rail Commuting Easier in Mumbai" (http://pib.nic.in/feature/feyr2001/fsep2001/f240920011.html). Press release. . Retrieved 2009-08-29.

[198] "Overview of existing Mumbai suburban railway" (http://www.mrvc.indianrail.gov.in/overview.htm). Mumbai Rail Vikas Corporation. . Retrieved 2008-07-07.

[199] Environment and urbanization 2002, p. 160 (http://books.google.co.in/books?id=0DBhYWmqpDoC&pg=PA160&dq=over+congestion+suburban+rail+mumbai#v=onepage&q=over congestion suburban rail mumbai&f=false)

[200] "Mumbai Metro Rail Project" (http://www.mmrdamumbai.org/projects_metro_rail.htm). Mumbai Metropolitan Region Development Authority (MMRDA). . Retrieved 2009-06-14.

[201] "Mumbai monorail to run in two years" (http://timesofindia.indiatimes.com/articleshow/2413046.cms). The Times of India. 2007-09-27. . Retrieved 2009-03-19.

[202] "Terminal Facilities in Metropolitanc Cities" (http://164.100.24.208/ls/CommitteeR/Railways/21streport.pdf) (PDF). Ministry of Railways (India). p. 14. . Retrieved 2009-08-28. "The port city of Mumbai is served by 5 passenger terminals namely Chhatrapati Shivaji Terminal (CST), Mumbai Central, Dadar, Bandra and Lokmanya Tilak Terminal."

[203] http://www.thaindian.com/newsportal/business/mumbai-airports-traffic-control-tower-design-bags-award_100221024.html

[204] "Chhatrapati Shivaji International Airport (CSIA)" (http://www.csia.in/masterplan.asp). Csia.in. . Retrieved 2010-05-05.

[205] "Work on Navi Mumbai airport may start next year" (http://www.thehindubusinessline.com/2006/12/19/stories/2006121901370700.htm). *Business Line* (The Hindu). 2006-12-19. . Retrieved 2009-05-16.

[206] "MIAL eyes Juhu airport" (http://www.mid-day.com/news/2007/jun/194964.htm). MiD DAY. 2007-06-07. . Retrieved 2009-06-14.

[207] Executive Summary on Comprehensive Transportation Study for MMR, p. 2–12

[208] Chittar 1973, p. 65: "The Port is endowed with one of the best natural harbours in the world and has extensive wet and dry dock accommodation to meet the normal needs of the city."

[209] Press Information Bureau (Government of India) (2003-01-07). "Laudable Achievement of JNPT" (http://pib.nic.in/archieve/lreleng/lyr2003/rjan2003/07012003/r070120037.html). Press release. . Retrieved 2009-08-29.

[210] "Our Mission" (http://www.google.co.in/search?hl=en&safe=off&q=site:jnport.com+55-60%&btnG=Search&meta=&aq=f&oq=). Jawaharlal Nehru Port Trust. . Retrieved 2009-08-27.

[211] Sonawane, Rakshit (2007-05-13). "Cruise terminal plan gets MoU push" (http://cities.expressindia.com/fullstory.php?newsid=236291). *Daily News and Analysis*. . Retrieved 2009-08-27. "While Arthur Bunder is used by small boats and Hay Bunder caters to declining traffic of barges, Ferry Wharf offers services to Mora, Mandva, Rewas and Uran ports."

[212] "BMC Inc. will now sell bottled water" (http://www.indianexpress.com/ie/daily/19980521/14150784.html). *Express News Service* (The Indian Express). 1998-05-21. . Retrieved 2009-06-13.

[213] Sawant, Sanjay (2007-03-23). "It will be years before Mumbai surmounts its water crisis" (http://www.dnaindia.com/mumbai/report_it-will-be-years-before-mumbai-surmounts-its-water-crisis_1086577). *Daily News and Analysis (DNA)*. . Retrieved 2009-06-13.

[214] "Tansa water mains to be replaced" (http://timesofindia.indiatimes.com/articleshow/msid-2247432,prtpage-1.cms). The Times of India. 2007-08-01. . Retrieved 2009-06-13.

[215] Wajihuddin, Mohammed (2003-05-04). "Make way for Mulund, Mumbai's newest hotspot" (http://cities.expressindia.com/fullstory.php?newsid=50939). *Mumbai Newsline* (Indian Express Group). . Retrieved 2009-06-13.

[216] "Country's first water tunnel to come up in Mumbai" (http://www.dnaindia.com/report.asp?NewsID=1151960). *Daily News and Analysis (DNA)*. 2008-02-20. . Retrieved 2008-02-21.

[217] Express News Service (22 October 2009). "Now, a toll-free helpline to check water leakage, theft" (http://in.news.yahoo.com/48/20091022/804/tnl-now-a-toll-free-helpline-to-check-wa.html). Yahoo India News. . Retrieved 2009-10-22.

[218] Nevin, John (2005-08-27). "Plastic ban: 1 lakh to be jobless" (http://www.rediff.com/money/2005/aug/27plastic.htm). Rediff. . Retrieved 2009-06-13.

[219] "How BMC cleans up the city" (http://www.mid-day.com/news/2002/aug/29797.htm). MiD DAY. 2002-08-26. . Retrieved 2009-06-13.

[220] "Bombay Sewage Disposal" (http://www.worldbank.org.in/external/default/main?pagePK=64027221&piPK=64027220&theSitePK=295584&menuPK=295621&Projectid=P010480). The World Bank Group. . Retrieved 2009-05-12.

[221] Dasgupta, Devraj (2007-04-26). "Stay in island city, do biz" (http://timesofindia.indiatimes.com/Cities/Mumbai/Stay_in_island_city_do_biz/articleshow/1956009.cms). The Times of India. . Retrieved 2009-06-13.

[222] "NTPC to give Mumbai 350 mw; electricity tariff may go up" (http://www.financialexpress.com/news/ntpc-to-give-mumbai-350-mw-electricity-tariff-may-go-up/181350/). *The Financial Express* (Indian Express Group). 2006-10-21. . Retrieved 2009-06-13.

[223] Campbell 2008, p. 143

[224] Somayaji, Chitra; Bhatnagar, Shailendra (2009-06-13). "Reliance Offers BlackBerry in India, Vies With Bharti" (http://www.bloomberg.com/apps/news?pid=20601082&sid=a_3Aeo82P.Kg&refer=canada). Bloomberg. . Retrieved 2009-06-13.

[225] "MTNL Launches IPTV Services On Broadband" (http://mumbai.mtnl.net.in/triband/voip/faq.html). *MTNL TriBand, Mumbai*. . Retrieved 2009-05-13.

[226] "Broadband &Internet" (http://www.airtel.in/wps/wcm/connect/airtel.in/Airtel.In/Home/ForYou/Broadband+Internet/). Airtel. . Retrieved 2009-05-13.

[227] Population and Employement profile of Mumbai Metropolitan Region, p. 6

[228] Population and Employement profile of Mumbai Metropolitan Region, p. 13

[229] "India: largest cities and towns and statistics of their population" (http://www.world-gazetteer.com/wg.php?x=&men=gcis&lng=en&dat=80&geo=-104&srt=pnan&col=aohdq&msz=1500&pt=c&va=&srt=pnan). World Gazetteer. . Retrieved 2008-01-31.

[230] "India: metropolitan areas" (http://www.world-gazetteer.com/wg.php?x=&men=gcis&lng=en&dat=80&geo=-104&srt=pnan&col=aohdq&msz=1500&va=&pt=a). World Gazetteer. . Retrieved 2008-01-17.

[231] Population and Employement profile of Mumbai Metropolitan Region, p. 12

[232] "Number of Literates & Literacy Rate" (http://censusindia.gov.in/Census_Data_2001/India_at_glance/literates1.aspx). *Census Data 2001: India at a Glance*. Registrar General & Census Commissioner, India. . Retrieved 2009-04-26.

[233] "Sex Ratio" (http://censusindia.gov.in/Census_Data_2001/India_at_glance/fsex.aspx). *Census Data 2001: India at a Glance*. Registrar General & Census Commissioner, India. . Retrieved 2009-04-26.

[234] "Parsis top literacy, sex-ratio charts in city" (http://timesofindia.indiatimes.com/articleshow/843036.cms). The Times of India. 2004-09-08. . Retrieved 2009-07-02.

[235] "11th annual report Crime in Maharashtra 2008:Criminal Investigation Department, Pune" (http://docs.google.com/viewer?url=http://mahacid.com/cid/Preface.pdf). Docs.google.com. . Retrieved 2010-05-05.

[236] "Census GIS Household" (http://www.censusindiamaps.net/page/Religion_WhizMap1/housemap.htm). *Census of India*. Office of the Registrar General. . Retrieved 2008-12-09.

[237] Mehta 2004, p. 99: "Maharashtrians now comprise 60 percent of the city's residents; 19 percent are Gujarati, and the rest are Muslim, North Indian, Sindhi, South Indian, Christian, Sikh, Parsi, and everybody else."

[238] Bates 2003, p. 266

[239] Baptista 1967, p. 5

[240] Strizower 1971, p. 15

[241] " The world's successful diasporas (http://www.managementtoday.co.uk/news/648273/)". Managementtoday.co.uk.

[242] Pai 2005, p. 1804

[243] O'Brien 2003, p. 141

[244] Datta & Jones 1999, Low-Income Households and the Housing Problem in Mumbai, pp. 158–159

[245] "Slum Cities: A Shifting World" (http://www.cbc.ca/correspondent/060507.html). *CBC News* (Canadian Broadcasting Corporation). 2006-05-07. . Retrieved 2008-08-16.

[246] Jacobson, Marc (May 2007). "Dharavi: Mumbai's Shadow City" (http://ngm.nationalgeographic.com/2007/05/dharavi-mumbai-slum/jacobson-text). *National Geographic*. National Geographic Society. . Retrieved 2009-04-28.

[247] Davis 2006, p. 31

[248] "Highlights of Economic Survey of Maharashtra 2005-06" (http://maccia.org.in/ecoSmaha06.pdf) (PDF). Directorate of Economics and Statistics, Planning Department (Government of Maharashtra). p. 2. . Retrieved 2008-02-13.

[249] National Crime Records Bureau (2007). "Crime in India-2007" (http://ncrb.nic.in/cii2007/home.htm) (PDF). Ministry of Home Affairs (Government of India). p. 2. . Retrieved 2009-04-25.

[250] "Once again, Arthur Road Jail prepares for mother of all trials" (http://www.indianexpress.com/news/once-again-arthur-road-jail-prepares-for-mo/412831/). The Indian Express. 2009-01-20. . Retrieved 2009-09-06.

[251] "Principal Cities" (http://www.maharashtra.gov.in/english/community/community_citiesShow.php). Government of Maharashtra. . Retrieved 2009-07-07.

[252] "Beginners' Bollywood" (http://www.theage.com.au/news/arts/beginners-bollywood/2005/09/27/1127804471297.html). Sydney: The Age. 2005-09-28. . Retrieved 2008-09-11.

[253] Vilanilam 2005, p. 130 (http://books.google.com/books?id=kvQtoquTT6kC&printsec=frontcover#PPA130,M1)

[254] Huda 2004, p. 203 (http://books.google.com/books?id=JIwxhDGO-mUC&pg=PA203&dq=Mumbai+world's+largest+IMAX+dome+theatre)

[255] Nagarajan, Saraswathy (2006-09-10). "Matchbox journeys" (http://www.hindu.com/mag/2006/09/10/stories/2006091000210500.htm). *The Hindu*. . Retrieved 2009-06-11.

[256] "Filmfare Awards gets new sponsor" (http://movies.indiatimes.com/articleshow/msid-1367349,prtpage-1.cms). *IndiaTimes Movies* (The Times of India). 2006-01-11. . Retrieved 2008-09-11.

[257] Chaudhuri 2005, pp. 4–6

[258] Gilder, Rosamond (October 1957). "The New Theatre in India: An Impression". *Educational Theatre Journal* (Washington, DC: Johns Hopkins University Press) **9** (3): 201–204. ISSN 0192-2882.

[259] David 1995, p. 232

[260] "Chhatrapati Shivaji Maharaj Vastu Sangrahalaya" (http://www.bombaymuseum.org/). Prince of Wales Museum of Western India, Mumbai. . Retrieved 2007-01-30.

[261] Sharma, Archana (2003-10-13). "Jijamata Udyan: A zoo without a view" (http://timesofindia.indiatimes.com/cms.dll/html/uncomp/articleshow?msid=230113). The Times of India. . Retrieved 2009-05-13.

[262] "Sahitya Akademi: awards and fellowships" (http://www.sahitya-akademi.gov.in/old_version/awa1.htm). Sahitya Akademi. 1999. . Retrieved 2009-07-08.

[263] Bavadam, Lyla (22 June – 5 July 2002). "Mumbai, past in present" (http://www.hinduonnet.com/fline/fl1913/19130670.htm). *Frontline* **19** (13). . Retrieved 2009-07-07.

[264] "Rainswept glory" (http://www.hindu.com/mag/2004/07/25/stories/2004072500330200.htm). *The Hindu*. 2004-07-24. . Retrieved 2009-07-07.

[265] Morris & Winchester 2005, p. 212

[266] "Mumbai's entrance -the 'Gateway' to be more tourist-friendly" (http://www.hindu.com/thehindu/holnus/002200703041014.htm). *The Hindu*. 2007-03-04. . Retrieved 2009-07-07.

[267] Bavadam, Lyla (April 11–24, 2009). "Forgotten classics" (http://www.hinduonnet.com/fline/fl2608/stories/20090424260806600.htm). *Frontline* **26** (08). . Retrieved 2009-07-07.

[268] "Tall Buildings of Mumbai" (http://www.emporis.com/en/wm/ci/bu/sk/?id=102037). Emporis. . Retrieved 15 August 2009.

[269] "India: World heritage sites centre" (http://whc.unesco.org/en/statesparties/in). UNESCO. . Retrieved 2007-08-09.

[270] "About Essel World" (http://www.esselworld.com/). Essel World. . Retrieved 2008-01-29.

[271] O'Brien 2003, p. 143

[272] "Kala Ghoda Arts Festival" (http://www.kalaghodaassociation.com/). Kala Ghoda Association. . Retrieved 2008-02-06.

[273] Shah, Shika (2008-09-17). "Bandra's spirit captured in cakes, tattoos" (http://www.mid-day.com/whatson/2008/sep/170908-Bandra-Fair-carnival-best-buys.htm). *MiD DAY*. . Retrieved 2008-09-27.

[274] "The Banganga Festival" (http://www.maharashtratourism.gov.in/MTDC/HTML/MaharashtraTourism/Default.aspx?strpage=../MaharashtraTourism/MTDC_Festival/Banganga_Festival.html). Maharashtra Tourism Development Corporation. . Retrieved 2008-02-07.

[275] "The Elephanta Festival" (http://www.maharashtratourism.gov.in/MTDC/HTML/MaharashtraTourism/Default.aspx?strpage=../MaharashtraTourism/MTDC_Festival/Elephanta_Festival.html). Maharashtra Tourism Development Corporation. . Retrieved 2008-02-07.

[276] "Mumbai celebrates Maharashtra Day" (http://timesofindia.indiatimes.com/Cities/Mumbai-celebrates-Maharashtra-Day/articleshow/4471265.cms). The Times of India. 2009-05-01. . Retrieved 2009-07-06.

[277] Krishnan, Ananth (2009-03-24). "'Vote at Eight' campaign" (http://www.hindu.com/2009/03/24/stories/2009032450711300.htm). The Hindu. . Retrieved 2009-07-06.

[278] "Sister Cities of Los Angeles" (http://www.lacity.org/SisterCities/). Official Website of the City of Los Angeles. . Retrieved 2008-02-08.

[279] ""Stuttgart Meets Mumbai": 40th Anniversary Celebrations of the Sister City Relationship" (http://www.indianembassy.de/template.php?mnid=104&inclpage=stuttgartmeeting.htm). The Embassy of India, Berlin. . Retrieved 2008-02-08.

[280] "Yokohama of the World" (http://www.city.yokohama.jp/ne/info/map/worldE.html). City of Yokohama. . Retrieved 2008-02-08.

[281] "Eight Cities/Six Ports: Yokohama's Sister Cities/Sister Ports" (http://www.welcome.city.yokohama.jp/eng/tourism/mame/a3000.html). Yokohama Convention & Visitiors Bureau (http://www.welcome.city.yokohama.jp/eng/ycvb/index.html). . Retrieved 2009-07-18.

[282] Bansal, Shuchi; Mathai, Palakunnathu G. (2005-04-06). "Mumbai's media Mahabharat" (http://www.rediff.com/cms/print.jsp?docpath=//money/2005/apr/06spec1.htm). Rediff. . Retrieved 2009-05-14.

[283] Rao, Subha J. (2004-10-16). "Learn with newspapers" (http://www.hindu.com/yw/2004/10/16/stories/2004101600260300.htm). *The Hindu*. . Retrieved 2009-05-14.

[284] Naregal, Veena (2002-02-05). "Privatising emancipation (A Book Review)" (http://www.hinduonnet.com/thehindu/br/2002/02/05/stories/2002020500040500.htm). *Language, Politics, Elites, and the Public Sphere* (The Hindu). . Retrieved 2007-12-24.

[285] "DD Mumbai bid to boost revenue" (http://www.thehindubusinessline.com/businessline/2000/07/01/stories/190102dd.htm). *Business Line* (The Hindu). 2000-07-01. . Retrieved 2009-06-10.

[286] "IN-fighting among cable operators" (http://www.indianexpress.com/res/web/pIe/ie/daily/19990826/ige26046.html). The Indian Express. 1999-07-26. . Retrieved 2009-06-10.

[287] "What is CAS? What is DTH?" (http://www.rediff.com/money/2006/sep/05iycu.htm). *Rediff News*. Rediff. 2006-09-05. . Retrieved 2009-06-10.

[288] "Tata Sky on Insat 4A" (http://www.lyngsat.com/packages/tatasky.html). *LyngSat*. . Retrieved 2008-08-10.

[289] "Radio stations in Maharashtra, India" (http://www.asiawaves.net/india/maharashtra-radio.htm#mumbai-radio). Asiawaves. . Retrieved 2008-01-18.

[290] D.S., Madhumathi (2001-12-29). "WorldSpace sees big gains in the long run" (http://www.thehindubusinessline.com/demo/stories/2001122900810200.htm). *Business Line* (The Hindu). . Retrieved 2009-06-10.

[291] "Few takers for CAS in Mumbai" (http://timesofindia.indiatimes.com/articleshow/856609.cms). The Times of India. 2006-12-20. . Retrieved 2008-01-22.

[292] Ganti 2004, p. 3

[293] Lundgren, Kari (2008-11-26). "Bollywood Trawls London for Talent as Students Balk at Banking" (http://www.bloomberg.com/apps/news?pid=20601085&sid=a6JQLLVfORXM&refer=europe). Bloomberg. . Retrieved 2009-05-26.

[294] "Bollywood filmmakers experimenting with new genre of films" (http://economictimes.indiatimes.com/articleshow/3373564.cms). *The Economic Times* (The Times of India). 2008-07-17. . Retrieved 2009-06-10.

[295] Deshpande, Haima (2001-03-05). "Mumbai's Film City may be home to world cinema" (http://www.indianexpress.com/ie/daily/20010305/ina05024.html). The Indian Express. . Retrieved 2009-05-14.

[296] Gupta 2006, p. 70 (http://books.google.co.in/books?id=_aoH81IN8SsC&pg=PA70&dq=Marathi+film+industry+based+in+Mumbai&lr=)

[297] "City has 43 one-teacher schools" (http://www.mid-day.com/news/2006/sep/144108.htm). *MiD DAY* (MiD-Day Infomedia). 2006-09-24. . Retrieved 2009-06-09.

[298] Mukherji, Anahita (2009-04-02). "Education board tells schools to get state recognition" (http://timesofindia.indiatimes.com/articleshow/4346890.cms). The Times of India. . Retrieved 2009-06-09.

[299] "Now, schools can teach in 2 languages" (http://timesofindia.indiatimes.com/articleshow/1516877.cms). The Times of India. 2006-05-05. . Retrieved 2009-06-09.

[300] Kak, Subhash (2004-07-13). "Saving India through Its Schools" (http://www.rediff.com/news/2004/jul/13kak.htm). *Rediff News*. Rediff. . Retrieved 2009-05-13.

[301] "Are you cut out for Arts, Science or Commerce?" (http://www.rediff.com/getahead/2008/jun/19trans.htm). *Rediff News* (Rediff). 2008-06-19. . Retrieved 2009-06-09.

[302] Sharma, Archana (2004-06-04). "When it comes to courses, MU dishes up a big buffet" (http://timesofindia.indiatimes.com/articleshow/718303.cms). The Times of India. . Retrieved 2009-06-09.

[303] "History" (http://www.mu.ac.in/History.html). University of Mumbai. . Retrieved 2009-06-09.

[304] "IIT flights return home" (http://www.dnaindia.com/report.asp?NewsID=1070723). *Daily News and Analysis (DNA)*. 2006-12-22. . Retrieved 2009-06-09.

[305] "About the Institute" (http://www.vjti.ac.in/home_about.asp). Veermata Jijabai Technological Institute (VJTI),. . Retrieved 2009-06-09.

[306] "Admission process for autonomous engg colleges to start today" (http://www.expressindia.com/latest-news/admission-process-for-autonomous-engg-colleges-to-start-today/321286/). Indian Express Group. 2008-06-11. . Retrieved 2009-06-09.

[307] "About University" (http://sndt.digitaluniversity.ac/Content.aspx?ID=7&ParentMenuID=7). SNDT Women's University. . Retrieved 2009-06-09.

[308] Bansal, Rashmi (2004-11-08). "Is the 'IIM' brand invincible?" (http://in.rediff.com/getahead/2004/nov/08rash.htm). *Rediff News* (Rediff). . Retrieved 2009-06-09.

[309] "Sydenham College: Our Profile" (http://www.sydenham.edu/our_profile.html). Sydenham College. . Retrieved 2009-04-26.

[310] "About The Government Law College" (http://www.glc.edu/incept.asp). Government Law College. . Retrieved 2009-04-26.

[311] Martyris, Nina (2002-10-06). "JJ School seeks help from new friends" (http://timesofindia.indiatimes.com/articleshow/24305727.cms). *The Times of India*. . Retrieved 2009-05-13.

[312] "University ties up with renowned institutes" (http://www.dnaindia.com/report.asp?NewsID=1065998). *Daily News and Analysis (DNA)*. 2006-11-24. . Retrieved 2009-06-09.

[313] "CIRUS reactor" (http://www.barc.ernet.in/webpages/reactors/cirus.html). Bhabha Atomic Research Centre (BARC). . Retrieved 2009-05-12.

[314] "About BCCI" (http://www.bcci.tv/about-bcci.html). Board of Control for Cricket in India (BCCI). . Retrieved 2009-05-16.

[315] "I-T Raids at IPL Headquarter at BCCI in Mumbai, reports NDTV | InvestmentKit.com Articles" (http://www.investmentkit.com/latestnews/2010/04/15/i-t-raids-at-ipl-headquarter-at-bcci-in-mumbai-reports-ndtv-2/). Investmentkit.com. 2010-04-15. . Retrieved 2010-05-05.

[316] Makarand, Waingankar (2009-01-18). "Attacking pattern of play has delivered" (http://www.hindu.com/2009/01/18/stories/2009011856451500.htm). The Hindu. . Retrieved 2009-06-08.

[317] Seth, Ramesh (2006-12-01). "Brabourne — the stadium with a difference" (http://www.hindu.com/yw/2006/12/01/stories/2006120100150200.htm). The Hindu. . Retrieved 2009-06-08.

[318] "Aussies claim elusive trophy" (http://www.smh.com.au/news/cricket/aussies-claim-elusive-trophy/2006/11/06/1162661575823.html). The Sydney Morning Herald. . Retrieved 2009-06-18.

[319] Srivastava, Sanjeev (2002-11-05). "Tendulkar serves it up" (http://news.bbc.co.uk/2/low/south_asia/2404371.stm). *BBC News* (BBC). . Retrieved 2009-06-08.
[320] Murali, Kanta (August-September 2002). "Gavaskar: India's Greatest Crickter" (http://www.hinduonnet.com/fline/fl1918/19180820.htm). *Frontline* (The Hindu) **19** (18). . Retrieved 2009-04-25.
[321] Bubna, Shriya (2006-07-07). "Forget cricket, soccer's new media favourite" (http://www.rediff.com/money/2006/jul/07fifa.htm). *Rediff News* (Rediff). . Retrieved 2009-06-09.
[322] "Mumbai Football Club launched" (http://www.rediff.com/sports/2007/jun/28foot.htm). *Rediff News* (Rediff). 2007-06-28. . Retrieved 2009-06-09.
[323] Sharma, Amitabha Das (2003-07-07). "Mahindra United in summit clash" (http://www.hindu.com/thehindu/2003/08/07/stories/2003080705022100.htm). The Hindu. . Retrieved 2009-06-09.
[324] "I-League: Mahindra United to face Mumbai FC" (http://www.hindu.com/thehindu/holnus/007200810101668.htm). The Hindu. 2008-10-10. . Retrieved 2009-06-09.
[325] "Stage set for Premier Hockey League" (http://www.rediff.com/sports/2004/nov/17hock.htm). *Rediff News* (Rediff). 2004-11-17. . Retrieved 2009-06-09.
[326] Pal, Abir (2007-01-17). "Mallya, Diageo fight for McDowell Derby" (http://timesofindia.indiatimes.com/articleshow/1233374.cms). The Times of India. . Retrieved 2009-06-08.
[327] Pinto, Ashwin (2005-03-05). "ESS plans marketing blitz around F1" (http://www.indiantelevision.com/mam/headlines/y2k5/mar/marmam27.htm). Indiantelevision.com. . Retrieved 2009-04-26.
[328] "Motor racing-Force India F1 team to launch 2008 car in Mumbai" (http://uk.reuters.com/article/motorSportsNews/idUKL2521523620080125). *Thomson Reuters* (Reuters UK). 2008-01-25. . Retrieved 2008-01-27.
[329] Jore, Dharmendra (2004-11-14). "Formula 1 powerboating swooshes into Mumbai, tourism hope for city" (http://cities.expressindia.com/fullstory.php?newsid=68125). *Mumbai Newsline* (The Indian Express). . Retrieved 2009-05-14.
[330] "Mumbai marathon draws all defending champions" (http://www.earthtimes.org/articles/show/160916.html). *The Earth Times*. 2007-12-18. . Retrieved 2009-05-28.
[331] "Bangalore replaces Mumbai on ATP Tour circuit" (http://www.sportsline.com/tennis/story/10834314). *CBS Sports*. 2008-05-20. . Retrieved 2009-05-28.
[332] http://books.google.com/books?id=-jxGSsKqfHgC&pg=PA10-IA6&dq=mumbai+theater+movement&client=firefox-a
[333] http://books.google.com/books?id=-jxGSsKqfHgC&dq=Mahesh+Dattani++By+Asha+Kuthari+Chaudhuri&source=gbs_summary_s&cad=0
[334] http://books.google.co.in/books?id=0DBhYWmqpDoC&printsec=frontcover&source=gbs_v2_summary_r&cad=0#v=onepage&q=&f=false
[335] http://www.mmrdamumbai.org/docs/escts.pdf
[336] http://books.google.com/books?id=GTEa93azj9EC&pg=PA3&dq=bollywood+produces+per+year&client=firefox-a#PPA3,M1
[337] http://books.google.co.in/books?id=_aoH81IN8SsC&printsec=frontcover
[338] http://books.google.co.in/books?id=-y3iNt0djbQC&printsec=frontcover&source=gbs_v2_summary_r&cad=0#v=onepage&q=&f=false
[339] http://books.google.com/books?id=JIwxhDGO-mUC&printsec=frontcover
[340] http://mdmu.maharashtra.gov.in/pages/Mumbai/mumbaiplanShow.php
[341] http://www.mumbaipolice.org/right_of_information/Right_of_Information.pdf
[342] http://www.lingref.com/isb/4/141ISB4.PDF
[343] http://www.mmrdamumbai.org/docs/Population%20and%20Employment%20profile%20of%20MMR.pdf
[344] http://stateelection.maharashtra.gov.in/pdf/THE%20MUMBAI%20MUNICIPAL%20CORPORATION%20ACT_1888.pdf
[345] http://ia311321.us.archive.org/3/items/shorthistoryofbo00coxerich/shorthistoryofbo00coxerich.pdf
[346] http://dsal.uchicago.edu/reference/gazetteer/toc.html?volume=8
[347] http://dsal.uchicago.edu/reference/gazetteer/
[348] http://www.mcgm.gov.in/

SS Fort Stikine

SS Fort Stikine	

Smoke billowing out of the site	
Date	14 April 1944
Time	16:15 IST (10:45 UTC)
Location	Victoria Dock, Bombay, British India
Casualties	
800 dead	
3,000 injured	

The **Bombay Explosion** (or Bombay Docks Explosion) occurred on 14 April 1944, in the Victoria Dock of Bombay (now Mumbai) when the SS *Fort Stikine* carrying a mixed cargo of cotton bales, gold, ammunition including around 1,400 tons of explosive caught fire and was destroyed in two giant blasts, scattering debris, sinking surrounding ships and killing around 800 people.

The vessel, the voyage and cargo

The SS *Fort Stikine* was a 7142 gross ton freighter built in 1942 in Prince Rupert, British Columbia, under a lend-lease agreement, and was named for Fort Stikine, a former outpost of the Hudson's Bay Company located at what is now Wrangell, Alaska.

SS *Fort Stikine*

Sailing from Birkenhead on 24 February via Gibraltar, Port Said and Karachi, she arrived at Bombay on 12 April.

The ship carried

- explosives
- munitions
- Spitfires
- raw cotton bales
- oil barrels
- timber
- scrap iron
- gold bullion in 12.73 kg bars valued at £1–2 million.

One officer described the cargo as "just about everything that will either burn or blow up". The vessel berthed and was still awaiting unloading on 14 April.

Incident

In the mid-afternoon around 14.00, the crew were alerted to a fire onboard. Burning somewhere in the No. 2 hold, the crew, dockside fire teams and fireboats were unable to extinguish the conflagration, despite pumping over 900 tons of water into the ship, or find the source due to the dense smoke.

The explosion

At 15:50 the order to abandon ship was given, and sixteen minutes later there was a great explosion, cutting the ship in two and breaking windows over 12 km away. The two explosions were powerful enough to be recorded by seismographs at the Colaba Observatory in the city. Around two square miles were ablaze in an 800-metre arc around the ship, eleven neighbouring vessels were sunk or sinking, and the emergency personnel at the site suffered heavy losses. Attempts to fight the fire were dealt a further blow when a second explosion from the ship swept the area at 16:34.

Aftermath

It took three days to bring the fire under control, and later 8,000 men toiled for seven months to remove around 500,000 tons of debris and bring the docks back into action. The official death toll was 740, including 476 military personnel, with around 1,800 people injured; unofficial tallies run much higher. In total, twenty-seven other vessels were sunk or damaged in both Victoria dock and the neighbouring Prince's Dock.

Aftermath of the explosion at the harbour

Many families lost all their belongings and were left with just the clothes on their back. The government took full responsibility for the disaster and monetary compensation was paid to citizens who made a claim for loss or damage to property.

During normal dredging operations carried out periodically to maintain the depth of the docking bays one or two gold bars were found intact sporadically as late as the 1970s and returned to the British government. Once in every few years, gold bricks are recovered from Mumbai harbour, reminding everyone of the great tragedy, even six decades after the incident. Mumbai Fire Brigade's headquarters at Byculla has a memorial built in the memory of numerous fire fighters who died during this explosion. Fire Safety Week is observed all over Maharashtra from 14 April to 21 April in memory of Fire fighters who died in this explosion

A piece of propeller that landed in St. Xaviers High School, some three kilometres from the docks.

See

Mumbai Fire Brigade

External links

- The Great Bombay Explosion by John Ennis [1]
- Ship Explosions: SS Fort Stikine, Page 7-10 [2]
- The Great Bombay Explosion [3] By Lawrence Wilson
- Bombay Harbour Explosion [4]
- The day it rained gold [5]
- Ships lost in the Bombay Explosion [6]

Geographical coordinates: 18°57′10″N 72°50′42″E

Prince Rupert, British Columbia

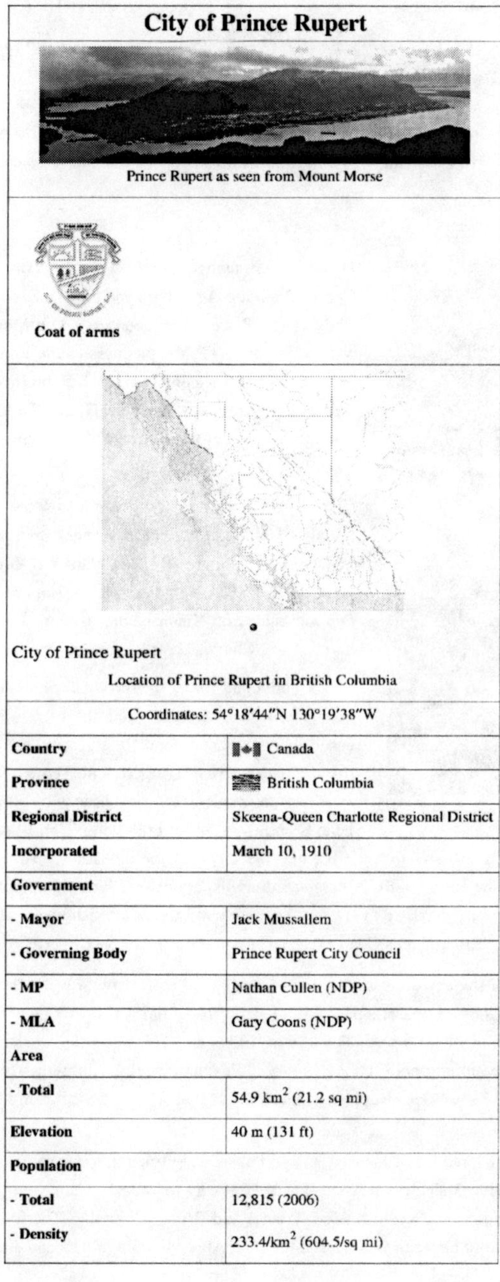

City of Prince Rupert	
Prince Rupert as seen from Mount Morse	
Coat of arms	
City of Prince Rupert Location of Prince Rupert in British Columbia	
Coordinates: 54°18′44″N 130°19′38″W	
Country	Canada
Province	British Columbia
Regional District	Skeena-Queen Charlotte Regional District
Incorporated	March 10, 1910
Government	
- Mayor	Jack Mussallem
- Governing Body	Prince Rupert City Council
- MP	Nathan Cullen (NDP)
- MLA	Gary Coons (NDP)
Area	
- Total	54.9 km^2 (21.2 sq mi)
Elevation	40 m (131 ft)
Population	
- Total	12,815 (2006)
- Density	233.4/km^2 (604.5/sq mi)

Time zone	Pacific Time Zone (UTC-8)
- Summer (DST)	Pacific Daylight Time (UTC-7)
Postal code span	V8J
Area code(s)	+1-250
Website	Prince Rupert.ca [1]

Prince Rupert is a port city in the province of British Columbia, Canada. It is the land, air, and water transportation hub of British Columbia's North Coast, and home to some 12,815 people (Statistics Canada, 2006).

History

Prince Rupert, May 1910. Looking north toward Mount Morse.

Prince Rupert, named after Prince Rupert of the Rhine, was founded by Charles Melville Hays, the general manager of the Grand Trunk Pacific Railway (GTP) and was incorporated on March 10, 1910. Prior to the opening of the GTP, the business centre on the North Coast was Port Essington on the Skeena River. After the founding of Prince Rupert at the western terminus for the Grand Trunk Pacific Railway, Port Essington returned to being a fishing community and is now a ghost town.

Charles Hays had many grand ideas for Prince Rupert including berthing facilities for large passenger ships and the development of a major tourism industry. These plans fell through when Charles Hays perished April 15, 1912 on the RMS *Titanic*. Mount Hays, the larger of two mountains on Kaien Island, is named in his honour, as is a local high school, Charles Hays Secondary School.

The former Capitol Theatre built in 1928.

Local politicians used the promise of a highway connected to the mainland as an incentive and the city grew over the next several decades. American troops finally completed the 100 mile stretch of road between Prince Rupert and Terrace during World War II to facilitate the movement of thousands of allied troops to the Aleutian Islands and the Pacific. Following World War II, the fishing industry, particularly salmon and halibut, and forestry became the city's major industries. Prince Rupert was the Halibut Capital of the World until the early 1980s. A long-standing dispute over fishing rights in the Dixon Entrance to the Hecate Straight (pronounced as "hekk-et") between American and Canadian fisherman lead to the formation of the 54-40 or Fight Society. The United States Coast Guard maintains a military base in nearby Ketchikan, Alaska.

Over the years, hundreds of students were said to have largely paid their way through school by working in the then lucrative fishing industry. Construction of a pulp mill began in 1947 and was operating by 1951. The construction of coal and grain shipping terminals followed. The 1960s, 1970s and 1980s saw the construction of many amenities including a civic centre, swimming pool, public library, golf course and performing arts centre (recently renamed "The Lester Centre of the Arts"). Prince Rupert had much to offer as it transitioned from a fishing and mill town to a small city.

In the 1990s, both the fishing and forest industries experienced a significant downturn in economic activity. In July, 1997, Canadian fishermen blockaded the Alaska Marine Highway ferry M/V *Malaspina*, keeping it in the port as a protest in the salmon fishing rights dispute between Alaska and British Columbia. The forest industry died when the soft wood lumber dispute arose between Canada and the U.S. After the pulp mill closed down, many people were out of a job, and a significant amount of top of the line machinery was left dormant. After reaching a peak of about 18,000 in the early 1990s, Prince Rupert's population began to decline as people left in search of work.

The period from 1996 to 2004 saw difficult times for Prince Rupert, including closure of the pulp mill, the burning down of a fish plant and a significant population decline. 2005 may be viewed as a critical turning point though. The announcement of the construction of a container port in April 2005, combined with new ownership of the pulp mill, the 2004 opening of a new cruise ship dock, the resurgence of coal and grain shipping, and the prospects of increased heavy industry and tourism foretell a bright future for the area.

On August 22, 1949, a magnitude 8.1 earthquake destroyed windows and buildings swung. See 1949 Queen Charlotte earthquake.

Geography

Orthographic projection centred over Prince Rupert

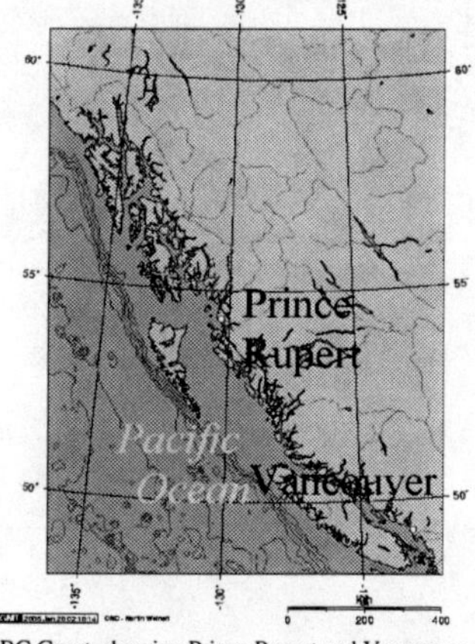

BC Coast, showing Prince Rupert and Vancouver

Prince Rupert is situated on Kaien Island (approximately 770 km (480 mi) north of Vancouver), just north of the mouth of Skeena River, and linked by a short bridge to the mainland. The city is located along the island's northwestern shore, fronting on Prince Rupert Harbour.

At the western terminus of Trans-Canada Highway 16 (the Yellowhead Highway), Prince Rupert is approximately 146 km (91 mi) west of Terrace, and 715 km (444 mi) west of Prince George.

Climate

Prince Rupert lies in a Marine West Coast climate (Koppen *Cfb*). It is very wet, with 2590 millimetres (102 in) of annual precipitation on average, 2470 millimetres (97.2 in) of that total being rain; in addition, 240 days per year have at least some precipitation, and there are only 1230 hours of sunshine per year. Summers are mild and comparatively drier, with an August high of 16.7 °C (62.1 °F). Spring and autumn are of normal length, but precipitation peaks in the autumn months. Winters are chilly and damp, but warmer than most locations at a similar latitude, due to Pacific moderation: the average January high is 4.6 °C (40.3 °F). Snow amounts are moderate for Canadian standards, averaging 126 centimetres (49.6 in) and occurring mostly from December to March.

Climate data for Prince Rupert

Month	Jan	Feb	Mar	Apr	May	Jun	Jul	Aug	Sep	Oct	Nov	Dec	Year
Record high °C (°F)	17.6 (64)	18.9 (66)	18.7 (66)	23.3 (74)	27.9 (82)	27.4 (81)	27.8 (82)	28.7 (84)	27 (81)	21.7 (71)	18.9 (66)	18.9 (66)	28.7 (84)
Average high °C (°F)	4.6 (40)	5.9 (43)	7.4 (45)	9.9 (50)	12.3 (54)	14.2 (58)	16.1 (61)	16.7 (62)	14.9 (59)	11.1 (52)	7.1 (45)	5.1 (41)	10.5 (51)
Average low °C (°F)	-2.1 (28)	-1.0 (30)	0.3 (33)	2.1 (36)	5.0 (41)	8.0 (46)	10.1 (50)	10.3 (51)	7.6 (46)	4.7 (40)	0.9 (34)	-0.8 (31)	3.8 (39)
Record low °C (°F)	-24.4 (-12)	-18.1 (-1)	-17.2 (1)	-7.1 (19)	-2.2 (28)	1.1 (34)	2.8 (37)	2.8 (37)	-2.2 (28)	-11.3 (12)	-20.6 (-5)	-22.8 (-9)	-24.4 (-12)
Precipitation mm (inches)	256.9 (10.11)	203.9 (8.03)	191.6 (7.54)	178.7 (7.04)	139.5 (5.49)	123.7 (4.87)	114.3 (4.5)	155.4 (6.12)	244 (9.61)	379.2 (14.93)	304.4 (11.98)	302 (11.89)	2593.6 (102.11)
Rainfall mm (inches)	217.7 (8.57)	179 (7.05)	174.4 (6.87)	173.3 (6.82)	139.5 (5.49)	123.7 (4.87)	114.3 (4.5)	155.4 (6.12)	244 (9.61)	378.9 (14.92)	293.7 (11.56)	274.7 (10.81)	2468.5 (97.19)
Snowfall cm (inches)	40.9 (16.1)	26.1 (10.3)	17.1 (6.7)	5.1 (2.0)	0 (0)	0 (0)	0 (0)	0 (0)	0 (0)	0.3 (0.1)	9.6 (3.8)	27.1 (10.7)	126.3 (49.7)
Sunshine hours	43.7	64.8	99.6	139.5	173.5	147.6	146.7	151.0	113.0	71.0	46.6	32.1	1229.1
Avg. rainy days	17.0	16.0	20.0	19.3	18.5	17.8	16.8	16.9	18.7	24.4	22.5	20.4	228.3
Avg. snowy days	7.3	5.2	4.6	1.8	0	0	0	0	0	0.2	2.9	5.5	27.5
Avg. precipitation days	20.8	18.7	21.7	19.7	18.5	17.9	16.8	16.9	18.7	24.3	23.1	22.6	239.7
Source: Environment Canada[2] *2009-07-11*													

Population

Statistics Canada has recorded the following population counts in their censuses. Census agglomerations are listed in parentheses.

- 2006 - 12,815 (13,392)
- 2001 - 14,643 (15,302)
- 1996 - 16,714 (17,414)
- 1991 - (17,359)
- Population by Age Group 2001
- Age Group = Population (% Distribution)
  - Under 18 years = 4,320 (28.2%)
  - 18 - 34 years = 3,370 (22.0%)
  - 35 - 54 years = 5,020 (32.8%)
  - 55 - 74 years = 2,075 (13.6%)
  - 75 years and over = 515 (3.4%)
  - Total - Age Groups = 15,300 (100.0%)
  - Median Age = 34.8
  - Source: BC Stats Population Estimates, 2004.

Government

The current mayor of Prince Rupert is Jack Mussallem. The current councillors of Prince Rupert are Nelson Kinney, Anna Ashley, Kathy Bedard, Gina Garon, Sheila Gordon-Payne, and Joy Thorkelson.

Prince Rupert is part of the Skeena—Bulkley Valley federal riding (electoral district). Nathan Cullen is the current Member of Parliament for the riding, and is a member of the New Democratic Party.

City Hall.

In the Legislative Assembly of British Columbia, Prince Rupert is a large portion of the North Coast riding. Gary Coons is the current Member of the Legislative Assembly. He is a member of the New Democratic Party of British Columbia. The NDP traditionally has strong support in the region.

Notable residents

After 1908, Thomas Dufferin "Duff" Pattullo became mayor of Prince Rupert. He went on to become the Premier of British Columbia from 1933-1941, as a member of the Liberal Party.

Two of the many totem poles in Prince Rupert are situated outside City Hall.

Alexander Malcolm Manson, the first lawyer in Prince Rupert, was elected to the BC Legislature in the riding of Omineca in 1916 as a Liberal. He became Speaker of the House in 1921 and the following year was appointed as both Attorney-General and Minister of Labour, serving in both capacities for six years. He was later appointed to the BC Supreme Court.

Iona Campagnolo began her political career when she was elected to Prince Rupert City Council in 1966. In 1974, she successfully ran for the Liberal Party in the federal riding of Skeena. In 1976 she was appointed Minister of Amateur Sports. She became president of the Liberal Party of Canada in 1982. She served as British Columbia's Lieutenant-Governor from 2001 to 2007.

In 1986, NDP candidate Dan Miller was elected to the Prince Rupert Electoral District and from August 1999 through February 2000 was Premier.

Rod Brind'Amour, captain of the NHL's Carolina Hurricanes

Paul Wong (Artist), Canadian Video Artist, now based in Vancouver, British Columbia.

Industry

Prince Rupert currently relies on the fishing industry, port, and tourism; however from 1951 to 2001 Prince Rupert also benefited from the Watson Island Pulp Mill, located less than 14 km (8 mi) outside of the city.

Transport

Seaport

Prince Rupert's sheltered harbour is the deepest ice-free natural harbour in North America, and the 3rd deepest natural harbour in the world. Situated at 54° North, the harbour is the northwestern most port in North America linked to the continent's railway network. Located on the Great Circle Route between eastern Asia and western North America, the port is the first inbound and last outbound port of call for cargo ships.

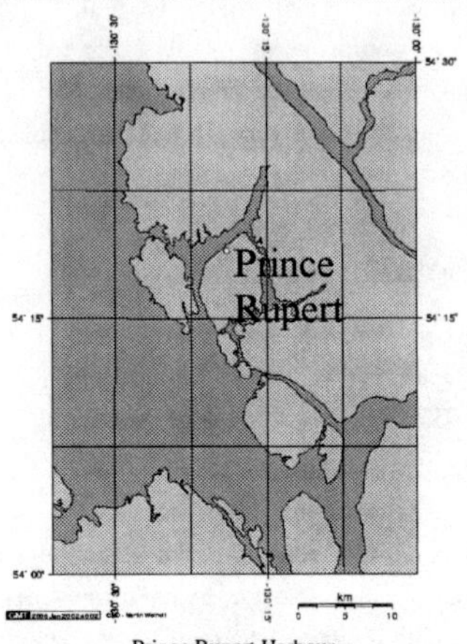

Prince Rupert Harbour

Passenger ferries operating from Prince Rupert include BC Ferries' service to the Queen Charlotte Islands and to Port Hardy on Vancouver Island, and Alaska Marine Highway ferries to Ketchikan, Juneau and Sitka and many other ports along Alaska's Inside Passage. The Prince Rupert Ferry Terminal is co-located with the Prince Rupert railway station, from which VIA Rail offers a thrice-weekly passenger train called *The Skeena*, connecting to Prince George and Jasper, and through a connection with *The Canadian* to the rest of the continental passenger rail network.

The Prince Rupert Port Authority is responsible for the port's operation.

Much of the harbour is formed by the shelter provided by Digby Island, which lies windward of the city and contains the Prince Rupert Airport. The city is located on Kaien Island and the harbour also includes Tuck Inlet, Morse Basin, Wainwright Basin, and Porpoise Harbour, as well as part of the waters of Chatham Sound which takes in Ridley Island.

Port Facilities

The Prince Rupert Port Authority (PRPA) is a federally appointed agency which administers and operates various port properties on the harbour. Previously run by the National Harbours Board and subsequently the Prince Rupert Port Corporation, the PRPA is now a locally run organization.

PRPA port facilities include:

- Atlin Terminal[3]
- Northlands Terminal[4]
- Lightening Dock
- Ocean Dock
- Westview Dock
- Fairview Terminal[5]
- Prince Rupert Grain[6]
- Ridley Terminals[7]
- Sulphur Corporation

All PRPA facilities are serviced by CN Rail.

The Canadian Coast Guard maintains CCG Base Seal Cove on Prince Rupert Harbour where vessels are homeported for search and rescue and maintenance of aids to navigation throughout the north coast. CCG also bases helicopters at Prince Rupert for servicing remote locations with aids to navigation, as well as operating a Marine Communications Centre, covering a large Vessel Traffic Services zone from Port Hardy at the northern tip of Vancouver Island to the International Boundary north of Prince Rupert.

Both BC Ferries and the Alaska Marine Highway operate ferries which call at Prince Rupert, with destinations in the Alaska Panhandle, the Queen Charlotte Islands, and isolated communities along the central coast to the south.

Airport

Prince Rupert Airport (YPR/CYPR) is located on Digby Island. Its position is 54°17′10″N 130°26′41″W, and its elevation is 35 m (116 ft[8]) above sea level. The airport consists of one runway, one passenger terminal, and two aircraft stands. Access to the airport is typically achieved by a bus connection that departs from one location in downtown Prince Rupert (Highliner Hotel) and travels to Digby Island by ferry. The airport is served by Air Canada and Hawkair from Vancouver International Airport (YVR).

Prince Rupert is also served by the Prince Rupert/Seal Cove Water Aerodrome, a seaplane facility with regularly scheduled, as well as chartered, flights to nearby villages and remote locations.

Railway

CN Rail has a mainline that runs to Prince Rupert from Valemount, British Columbia. At Valemount, the Prince Rupert mainline joins the CN mainline from Vancouver. Freight traffic on the Prince Rupert mainline consists primarily of grain, coal, wood products, chemicals, and as of 2007, containers. As the renovations at the Port of Prince Rupert continue, traffic on CN will steadily rise in future years.

In addition, a three times weekly passenger rail service known as *The Skeena* operated by Via Rail connects Prince Rupert with Prince George and Jasper. The service takes two days and requires an overnight hotel stay in Prince George. The route ends in Jasper and connects passengers with VIA's The Canadian, which runs between Toronto and Vancouver.

Weather

Prince Rupert is known as "The City of Rainbows", as it is Canada's wettest city, with an average annual precipitation of approximately 2,500 mm (100 in) (Statistics Canada, 1999). It is also regarded as the municipality in Canada which receives the least amount of sunshine annually. Winters are relatively mild for the latitude (even January does not average below freezing), although frosts and blasts of cold Arctic air from the northeast are not uncommon. Summers are relatively cool, with daytime temperatures averaging below 20°C (68°F). Wind speeds are relatively strong, with

Sunken Gardens near the courthouse.

prevailing winds blowing from the southeast. There is ample precipitation throughout the year, but autumn is the wettest season. Snowfall in Prince Rupert is rare and normally melts within a few days, although individual snowstorms may bring copious amounts of snow.

Tourist brochures boast about Prince Rupert's "100 days of sunshine".

Communications

Telephone, mobile, and Internet service are provided by CityWest (formerly CityTel). CityWest is owned by the City of Prince Rupert. CityWest provides long-distance telephone service, as does Telus.

In September 2005, the city changed CityTel from a city department into an independent corporation named CityWest. The new corporation immediately purchased the local cable company, Monarch Cablesystems, expanding CityWest's customer base to other northwest British Columbia communities.

Since January 2008, Rogers Communications has offered GSM and EDGE service in the area -- the first real competition to CityWest's virtual monopoly. Rogers offers local numbers based in Port Edward (prefix 600), which is in the local calling zone for the Prince Rupert area. The introduction of Rogers service forced Citywest to form a partnership with Bell Canada to bring digital services to Citywest Mobility, using CDMA.

Media

Radio

- **AM** 560 - CHTK, adult contemporary
- **AM** 860 - CFPR, CBC Radio One
- **FM** 100.7 - CIAJ, Christian programming
- **FM** 101.9 - CJFW-2, country music
- **FM** 98.1 - CFNR-FM, classic rock

Television

- Channel 6 - CFTK-1, CBC Television private affiliate

Tourist attractions

Prince Rupert is a central point on the Inside Passage, a route of relatively sheltered waters running along the Pacific coast from Vancouver, British Columbia to Skagway, Alaska. It is visited by many cruise ships during the summer en route between Alaska to the north and Vancouver and the Lower 48 to the south.

Prince Rupert is also the starting point for many wildlife viewing trips including whales, eagles and grizzly bears. The Khutzeymateen Grizzly Bear sanctuary features one of the densest remaining populations in North America; tours can be arranged by water or air (using float planes) departing from Prince Rupert.

Neighbouring communities

By virtue of location, Prince Rupert is the gateway to many destinations:

- Dodge Cove (1 km, 0.6 mi, west)
- Metlakatla (5 km, 3 mi, west)
- Port Edward (15 km, 9 mi, south)
- Lax Kw'alaams (Port Simpson) (30 km, 19 mi, northwest)
- Oona River (43 km, 27 mi, southwest)
- Kitkatla (65 km, 40 mi, south)
- Kisumkalum (140 km, 87 mi, east)
- Kitselas (142 km, 88 mi, east)
- Terrace (146 km, 87 mi, east)
- Hartley Bay (157 km, 98 mi, southeast)

The Queen Charlotte Islands (also known as Haida Gwaii) are to the west of Prince Rupert, across the Hecate Strait. Alaska is 49 nautical miles (90 km, 56 mi) north of Prince Rupert.

Citations

The book *Unmarked: Landscapes Along Highway 16*, written by Sarah de Leeuw, includes an essay about Prince Rupert entitled "Highway of Monsters".

Ra McGuire of the band Trooper wrote the hit song "Santa Maria" on a boat in Prince Rupert's Harbour. Says McGuire, "The boat was called The Lucky Lady. We sailed from Prince Rupert onto an island [Lucy Island[9]] off the coast with an awful lot of alcohol and some salmon to barbecue. Many of the lines in the song are direct quotes from the skipper. He actually said 'Okay, there's only fear and good judgment holding us back.' On the way back he said 'Does somebody know how to drive this thing?' I actually wrote these down in a little notepad as we went." [10]

See also

- Royal eponyms in Canada

External links

- City of Prince Rupert [1]
- Prince Rupert City and Regional Archives [11]
- Prince Rupert & District Chamber of Commerce [12]
- Skeena-Queen Charlotte Regional District [13]
- Prince Rupert Pictures [14]
- Prince Rupert Port Authority [15]
- Prince Rupert Airport [16]
- Northwest Community College (Prince Rupert Campus) [17]
- School District 52 (Prince Rupert) [18]
- Prince Rupert Secondary School [19]
- Charles Hays Secondary School [20]
- Prince Rupert Library [21]
- Prince Rupert Economic Development Corporation [22]
- Tourism Prince Rupert [23]

Geographical coordinates: 54°18′43.9″N 130°19′37.5″W

References

[1] http://www.princerupert.ca/
[2] Environment Canada— Canadian Climate Normals 1971–2000 (http://www.climate.weatheroffice.ec.gc.ca/climate_normals/results_e.html?Province=ALL&StationName=Prince Rupert&SearchType=BeginsWith&LocateBy=Province&Proximity=25&ProximityFrom=City&StationNumber=&IDType=MSC&CityName=&ParkName=&LatitudeDegrees=&LatitudeMinutes=&LongitudeDegrees=&LongitudeMinutes=&NormalsClass=A&SelNormals=&StnId=422&), accessed 11 July 2009
[3] http://www.rupertport.com/atlin.htm
[4] http://www.rupertport.com/northland.htm
[5] http://www.rupertport.com/container.htm
[6] http://www.rupertport.com/princerupertgrain.htm
[7] http://www.rupertport.com/ridleyterminals.htm
[8] This is a measured value in feet
[9] http://www.fogwhistle.ca/bclights/lucy/
[10] http://www.trooper.ca/default.php?cat=articles&subcat=35
[11] http://www.princerupertlibrary.ca/archives/
[12] http://www.princerupertchamber.ca/
[13] http://www.sqcrd.bc.ca/
[14] http://rupertpics.com/
[15] http://www.rupertport.com/
[16] http://www.ypr.ca/
[17] http://www.nwcc.bc.ca/campuses/rupert.cfm
[18] http://www.sd52.bc.ca/
[19] http://www.prss.net/
[20] http://charleshays.net/
[21] http://www.princerupertlibrary.ca/
[22] http://www.predc.com/
[23] http://www.tourismprincerupert.com/

Fort Stikine

Fort Stikine was a fur trade post and fortification in what is now the Alaska Panhandle, at the site of the present-day of Wrangell, Alaska, United States. Originally built as the **Redoubt San Dionisio** or **Redoubt Saint Dionysius** in 1834, the site was transferred to the British as part of a lease signed in the region in 1838, and renamed Fort Stikine when turned into a Hudson's Bay Company post in 1839. The post was closed and decommissioned by 1843 but the name remained for the large village of the Stikine people which had grown around it, becoming known as **Shakesville** in reference to its ruling Chief Shakes by the 1860s. With the Alaska Purchase of 1867, the fortification became occupied by the US Army and was re-named **Fort Wrangel**, a reference to Baron von Wrangel, who had been Governor of Russian America when the fort was founded. The site today is now part of the city of Wrangell.

History

Background

By a decree of Emperor Paul I known as the Ukase of 1799, the Russian Empire had in 1799 asserted ownership of the Pacific coast and adjoining lands of North America as far south as the 55th degree of latitude, with Novo-Arkhangelsk, today's Sitka, Alaska, founded shortly thereafter. In 1821, Emperor Alexander issued another ukase which extended the Russian claim south to 51 degrees north, also forbidding foreign vessels from approaching within 100 Italian miles of any Russian settlement.[1] Other powers protested and the line was withdrawn to "the line of the Emperor Paul", 55 degrees north, with parallel treaties with the United States and Great Britain (in 1824 and 1825 respectively) adjusting that southwards slightly to 54 degrees 40 minutes north so as to include all of Prince of Wales Island within Russian territory.

Redoubt Saint Dionysius

Under the Russo-British Treaty of 1825 (Treaty of St. Petersburg) establishing that boundary, and also establishing a land boundary northwards following the summit of the mountains, ten marine leagues from the coast, British rights to the Interior were guaranteed by Russia, along with the right of navigation of the Taku and Stikine Rivers, which were (and largely still are) the only access to those regions of what is now northern British Columbia. In 1833 Peter Skene Ogden went up the Stikine on reconnaissance and found the river too shallow for the HBC's sailing vessels. In 1834, under orders from John McLoughlin, Chief Factor of the Columbia Department of the HBC, based at Fort Vancouver, Ogden returned on the brig *Dryad* with the mission of establishing trade and a fort in the Stikine hinterland.[2] The expedition was met with armed opposition from the Russian-American Company, under Baron Wrangel's orders, near the river's mouth. During the ensuing confrontation and what would become a naval standoff, with the ships *Chichagof* and *Orel* dispatched from Sitka, Ogden and his men were driven off and Hudson's Bay Company stores, intended for trade and the establishment of the upriver post, were seized. Captain Lieutenant Dionysius Zarembo, commander of the *Chichagof*, under further orders, established a fortification on Zarembo Island which was named the Redoubt Saint Dionysius, also known as the Redoubt San Dionisio, to prevent further British attempts at access to the Stikine, control of which was seen to be contrary to the interests of the local fur trade on which Sitka depended.

Anglo-Russian Convention of 1838

When news of the confrontation reached Fort Vancouver, McLoughlin was outraged and quickly sent word to company HQ in London. In the ensuing diplomatic confrontation, resulting from Britain pressing indemnities on Russia for damages relating to the seizure and actions contrary to the treaty of 1825, Baron von Wrangel was forced out of office in disgrace because of the great cost in both money and prestige to the Empire. A treaty signed in 1838, known as the Anglo-Russian Convention, paid out a negotiated indemnity to Great Britain and also established the British right to build and maintain posts at the mouths of the Taku and Stikine as well as established a lease of the mainland and adjoining islands south from 56 degrees 30 minutes north, roughly the latitude of the Stikine. In return for this lease, Britain would supply so many furs per annum to the Russian American Company. Also included in the arrangement was an extension and expansion of shipments of vegetable, dairy and meat produce from the settlements of the HBC at Forts Langley, Nisqually and Vancouver (see Pugets Sound Agricultural Company).

Establishment of Fort Stikine

The following year, in 1839, James Douglas, later to be Chief Factor himself and also Governor of both Vancouver Island and British Columbia colonies, was sent north by McLoughlin to establish "Fort Stickeen" and what was formally called Fort Durham, but usually known as Fort Taku, or just Taku, under various spellings. Confrontations with the local group of Tlingit, the *Shtakeen Kwaan* ("Stikine Tribe") under Chief Shakes, near whose ceremonial clan house the fort had been erected, concderning control of the fur trade of the Stikine Country, led to the relocation of the principal village of that tribe to the location of the fort, and agreements with Shakes regarding control of trade in relation to other Tlingit groups and the inland Athapaskan peoples (the Tahltan, primarily) with whom the Stikines had long-standing agreements. Still, other tribes including the Haida, Nisga'a and Tsimshian traded at the post, with complicated consequences.

Fort Stikine and the slave trade

Much to the chagrin and horror of the company staff in charge of the post, the logistics of the fur trade resulted in an unexpected effect - an escalation of the slave trade by the Haida and Tlingits, with an accompanying rise in warfare and raiding against southern tribes (and each other) to provide goods for the purchase of furs to re-sell to the Hudson's Bay post. This, and the declining price of furs, and also the viability of the local supply, led to a decision by the company's Governor Simpson, after an 1841 visit, to close the northern posts, plus Fort McLoughlin farther south (near today's Bella Bella), to consolidate northern operations at Fort Simpson, located first near the mouth of the Nass, then at a more strategic location near today's Prince Rupert, off the mouth of the Skeena (today's Lax Kw'alaams).

Death of John McLoughlin, Jr.

The second Chief Trader appointed to Fort Stikine was the son of Chief Factor McLoughlin, John McLoughlin, Jr.. Unsuited to the appointment, the younger McLoughlin was unpopular with some of the Metis among the staff, who killed him in what was alleged by them to have been in self-defense at his drunken rage. Kanaka (Hawaiian) employees who witnessed the killing were to testify otherwise, and to allege that the rebel staff, led by one Urbain Heroux, had conspired with the local Tlingits to seize the post. Because the murder had happened on ostensibly Russian soil, the usual laws governing the Company and its staff, which were those of the Colony of Canada where the company's North American headquarters were, Heroux and the others were taken to Sitka for trial, and were ultimately released for lack of evidence. Governor Simpson, on his visit to the Russian American capital, was surprised to encounter Heroux at liberty on the streets of that town, but under Russian law the accused were free until convicted. Heroux was to live out his life in the Columbia District, but John Jr.'s death was said to be one of the factors embittering his father against the Company and his increasing sentiments and actions in favour of the American claim to what was becoming known as the Oregon Country.

Shakesville and Buck Choquette

In the company's absence, Chief Shakes took control of the post and of the Stikine River trade. Discovery of gold in the Queen Charlotte Islands in 1850, and then in the Thompson Country and Fraser Canyon in the later 1850s, led to wider encroachments and exploration by whites far beyond the locus of the Fraser Canyon Gold Rush of 1858-1861. One intrepid adventurer, Alexander "Buck" Choquette, originally from Quebec and who had been in the California goldfields, had already explored in the area of the Nass and other rivers between there and the Stikine. Equipped with an acquired proficiency in the Chinook Jargon, Choquette was at Fort Victoria when he met some of Shakes' people and persuaded them to bring him with them to the Stikine and what was left of Fort Stikine, which had by then become known as Shakesville (though still also referred to as Fort Stikine despite the absence of a formal post or a Chief Trader). Choquette was to earn the respect of Shakes and also the hand of his daughter Georgiana (or Georgie) as his wife, with the marriage consecrated according to the elaborate ceremonials of Tlingit custom.

Stikine Gold Rush

In the spring of 1861, Choquette set out on a canoe trip up the Stikine with his wife and ten warriors of the Stikines to prospect for gold, discovering it at what has been known as Buck Bar since, just southwest of Telegraph Creek (the gold-rush settlement at Telegraph Creek was, in fact, known as Buck's Bar until the construction era of the unfinished Collins Overland Telegraph). When word of Choquette's discovery reached the other goldfields and the colonial capitals, the Stikine Gold Rush was launched and hordes of men sought out the Stikine, with Fort Stikine aka Shakesville become an important port-of-call for steamboats now bound for the river's many gold-bearing bars. In response to the influx of miners, most of them (but not all) American, Governor Douglas decreed the creation of the Stikine Territory, covering the lands inland from Russian Ameriacn between the line of the Nass and Finlay Rivers north to the 62nd Parallel to prevent American feared American annexation of the region, just as he had created the Colony of British Columbia in similar circumstances and had witnessed the loss of the Oregon Country to American settlers previously.

Experienced from goldfields elsewhere, Choquette knew more money was to be made in provision of goods and supplies to the miner than in the workings themselves, and obtained rights to sell Hudson's Bay Company wares both at his upriver post and at a revived trading post at Shakesville.[3] Choquette was to maintain this post in an uneasy relationship with the Hudson's Bay Company, as well as his store upriver, which relocated at various times depending on fluctuations in the activity of the rush.

Although exploration and some mining continued, the rush was well over by 1867, when the United States purchased Alaska from the Russian Empire. Despite the profitability of American trade and a wider range of goods, Choquette was faced with the decision of retaining the Hudson's Bay Company license and the freedom from American taxation that came with it by moving to just within British territory, which is to say, ten marine leagues upriver (approximately 30 km). Choosing a site opposite the Great Glacier, Choquette opened another store near the confluence of what became known as the Choquette River near the Stikine Hot Springs (see also Boundary Range). In time, Choquette would be given charge of a customs post and Hudson's Bay outlet, though opted eventually to remain at his preferred location, which he named Ice Mountain (his name for the Great Glacier).

Fort Wrangel

With Choquette's departure, and the British flag with him, which had flown over his store at Shakesville, American troops took over the old fortification of Fort Stikine, renaming it Fort Wrangel. It was the second US Army post established in Alaska, the first being Fort Tongass on Tongass Island, immediately north of 54-40, but which was abandoned by 1870 as being of little real strategic or commercial value, as it was Fort Wrangel which controlled the main access inland and was therefore more viable as a customs port for the region, and Britain had shown no signs of military support for the claims that British Columbia had been making for its rights to the leased portion of the Panhandle, which had in any case been overtaken by American fishing, cannery and mining operations in the immediate aftermath of the Purchase.

Fort Wrangel again became a source of tension between American and British authorities in the region when more gold was discovered near Dease Lake in 1870, leading to the Cassiar Gold Rush of 1871. Once again thousands of miners poured into and through Fort Wrangel, and the US authorities attempted to exert control over British-registered shipping heading for the Stikine. A compromise was reached, and the confrontation derailed and prevented from escalating into warfare over the region. After the Cassiar rush was over, Fort Wrangel remained as one of the main US military installations in the region, and was again to play a strategic as well as a commercial role in relation to the Stikine's use as one of the lesser routes to the Klondike from 1897 and the mounting tensions of the Alaska Boundary Dispute, which was resolved by arbitration in 1903.

See also

- History of the west coast of North America
- Maritime Fur Trade

Further reading

- *The Dryad Affair: Corporate Warfare and Anglo-Russian Rivalry for the Alaskan Lisière*, J.W. Shelest, Conference Paper for "Borderlands", June 1989 [4]

  "This paper was originally presented at a 2-4 June 1989 conference dealing with the Yukon/Alaska/BC border and the issues surrounding this border held in Whitehorse, YT, Canada. The conference was jointly sponsored by the Yukon Historical & Museums Association (YHMA), Yukon College, The University of Victoria's Public History Group, and the Alaska Historical Society. The proceedings were published by the YHMA."

References

[1] Haycox, Stephen W. (2002). *Alaska: An American Colony* (http://books.google.com/books?id=8yu3pYpzLdUC). University of Washington Press. pp. 1118-1122. ISBN 9780295982496. .

[2] Mackie, Richard Somerset (1997). *Trading Beyond the Mountains: The British Fur Trade on the Pacific 1793-1843* (http://books.google.com/books?id=VKXgJw6K088C&pg=PA140). Vancouver: University of British Columbia (UBC) Press. p. 140. ISBN 0-7748-0613-3. .

[3] "Choquette River" (http://www.env.gov.bc.ca/bcgn-bin/bcg10?name=5922). BC Geographical Names Information System. .

[4] http://heritageyukon.ca/Shelest.html

Geographical coordinates: 56°28′15″N 132°22′36″W

Colaba Observatory

Colaba Observatory[1] [2] was an astronomical, timekeeping, geomagnetic and meteorological observatory located on the Island of Colaba, Mumbai (Bombay), India.

History

The Colaba Observatory [3] located in Bombay was built in 1826 by the East India Company for astronomical observations and time-keeping, with a purpose to provide support to British and other shipping which used Bombay as a port. The 165 year old Building served as office space for the Indian Institute of Geomagnetism. Geomagnetism and meteorological measurements were started here in 1841 by Arthur Bedford Orlebar, who was then Professor of Astronomy at Bombay's Elphinstone College. Magnetic measurements during the period 1841–mid 1845 were intermittent; following 1845 they became bi-hourly, then hourly. In 1845 Charles Brooke devised a self-recording photographic magnetometer[4] with a light-source, a mirror for amplifying the magnet's movement, and a drum of photographic paper. This system gradually replaced the older manual method of taking eye-observations through vertical microscope which scanned the ends of a magnetic needle; the magnetic needle itself was suspended by a bunch of silk fibers.

The new photographic drum method ensured continuous recording of geomagnetic elements, and rapidly gained use all over the world. It came to Colaba in 1871, when Charles Chambers (later to become F.R.S.) held the Directorship. Through his thorough examination of geomagnetic measurements at Colaba, and his masterly interpretation of the physics behind the phenomena, he brought great esteem to Colaba Observatory. After his untimely death in Feb. 1896, the mantle of Directorship fell on the shoulders of Dr. Nanabhoy Ardeshir Framji Moos, the first Indian to hold this position.

Dr. N. A. F. Moos was an able successor to Chambers. With an Engineering degree from Poona, and a higher degree in Science from Edinburgh in Scotland, he saw to the efficient functioning of Colaba Observatory, regular analysis and interpretation of the measurements, and the starting of seismological observations. In 1900, the existence of Colaba Observatory came under threat, when a decision was taken to electrify the fleet of horse drawn trams then existing in Bombay for public transport. The electric trams by generating electromagnetic noise, would have vitiated the data from the Colaba magnetic observatory.

The credit of selecting the alternate site of Alibag, located about 30 km to the south-east of Bombay as the crow flies goes to Dr. Moos. Alibag was located "far enough from Bombay to be free from the threatened electromagnetic noise, and yet near enough to retain the same geomagnetic characteristics"–these aspects were checked out carefully over a 2 year period 1904-1906, and then only was recording at Colaba discontinued, and the electric tram service started in Bombay. The entire building is made of hand-picked, non magnetic, Porbandar sandstone, and magnetic recording is carried on in a room built with such good insulation, that the variation in temperature within, even today, is just 10°C over an entire day.

Of the entire Colaba - Alibag data, the French geomagnetician Pierre Noel Mayaud, had the following to say in 1973:

"Finally, the (magnetic) records of Colaba and Alibag were found to form a beautiful series, beginning in 1871, and making up perhaps, the most complete collection of records in the world. Their quality and especially their regularity were particularly impressive, even in comparison with the Kew and Melbourne records".

Moos retired in 1919 after leading the Colaba-Alibag Observatories to worldwide renown. In 1910 he summarized the main findings from 50 years of geomagnetic measurement at the Colaba-Alibag Observatory over 1846-1905, in two volumes titled "Magnetic observations made at the Government Observatory, Bombay for the period 1846-1905. Parts I. and II." Of these volumes and of Colaba-Alibag's performance as a Geomagnetic Observatory, J.A. Fleming [5], a pioneer in Terrestrial Magnetism and Electricity, had the following to say in 1954:

"The Golden Jubilee of the foundation of the Magnetic Observatory at Alibag (Mumbai), is a historic one in the field of Geomagnetism, and marks the long established application of India in an unparalleled series of magnetic recording of the phenomena, and publication of interpretative discussions of the accumulated data, as prepared under the direction of India's foremost investigator (N.A.F. Moos) in the two large volumes.

Despite over 1500 selected references in the field of geomagnetic research, Volume 3 of the Physics-of-the-Earth Series of the United States National Research Council, there is none which exhibits so wide and varied and intensive coverage of all the geomagnetic problems in the early 20th century."

Of Moos, the Director who followed, Prof. K. R. Ramanathan (later to head the Physical Research Laboratory, Ahmedabad), said:

"He was an ideal head of the observatory, always taking a deep interest in the welfare of his staff, and being held by them in great affection and esteem".

Over the year 1919–1971, 17 Directors steered the Colaba-Alibag Observatories through avenues of meticulous and uninterrupted geomagnetic recordings, regular publishing of the data, and discussion of observations in scientific research journals.

In 1971, a momentous change occurred with the conversion of the Colaba-Alibag Observatories, into an autonomous research organisation called the Indian Institute of Geomagnetism. Till then the Colaba-Alibag Observatories were part of the Indian Meteorological Department. Its headquarters continued to be in Mumbai, in the sturdy 1826 building constructed by John Curin, Astronomer for the East India Company. The first Director of the Indian Institute of Geomagnetism over 1971–1979 was Prof. B. N. Bhargava. Prof. R. G. Rastogi was the next Director over 1980-1989.

During the IGY-IGC years of 1957–1959, Prof. K. R. Ramanathan (a past Director of Colaba-Alibag), strongly advocated the setting up of magnetic observatories to examine the equatorial electrojet. The TRIVANDRUM and ANNAMALAINAGAR observatories were set up in Nov. 1957, and were tended in their infancy, first under the Directorship of Mr. S. L. Malurkar, and then under Prof. P. R. Pisharoty.

Eighteen years elapsed before the need for further observatories along 75°E longitude meridian was felt, to serve the needs of the USSR sponsored "Project Geomagnetic Meridian". Ujjain and Jaipur were consequently set up in July 1975, as also Shillong at 92°E longitude. In May 1977, Gulmarg, located very near the focus of the Sq. current system was started. In May 1991, the ninth observatory was started at Nagpur and then the observatories Vishakhapatnam, Pondicherry and Tirunelveli followed. Apart from these, a temporary station was run in the Andaman Islands in 1974, as support for the ONGC (Oil and Natural Gas Commission of India) in petroleum prospecting. Since 1979, an array of Gough-Reitzel magnetometers has operated at various sites in India, for studies of the Earth's internal structure by examining electromagnetic induction within the earth. The Indian Institute of Geomagnetism thus operates Ten magnetic observatories.

- **Content provide with the permission of the Indian Institute of Geomagnetism.**

External links

- Magnetic Observations at the Colaba Observatory [6]
- The extreme magnetic storm of 1–2 September 1859 [7] by BT Tsurutani, WD Gonzalez, GS Lakhina, S Alex; JOURNAL OF GEOPHYSICAL RESEARCH, 2003 VOL. 108, NO. A7, 1268
- Research on Historical Records of Geomagnetic Storms [8] by G. S. Lakhina, S. Alex, B. T. Tsurutani, and W. D. Gonzalez; Proceedings IAU Symposium No. 226, 2005
- Google Books on the "Colaba Observatory" [9]
- Google Scholar on the "Colaba Observatory" [10]

References

[1] *Indian Institute of Geomagnetism* (http://iigs.iigm.res.in/history.htm)

[2] History of the Institute (http://iigm.res.in/iigweb/index.php/185)

[3] PART I. The Colaba Observatory and its Observations. (http://books.google.com/books?id=-A4EAAAAQAAJ&jtp=4) The Meteorology of the Bombay Presidency by Charles Chambers, F.R.S. (1878)

[4] On the Automatic Registration of Magnetometers, and other Meteorological Instruments, by Photography. By Charles Brooke, M.B., F.R.C.S., Pages 69-77 (http://books.google.com/books?id=iGwOAAAAIAAJ&pg=PA69) Philosophical Transactions of the Royal Society of London, Part 1 (1847)

[5] http://www.agu.org/inside/awards/fleming2.html

[6] http://adsabs.harvard.edu/abs/1911Natur..88..113.

[7] http://www.agu.org/pubs/crossref/2003/2002JA009504.shtml

[8] http://journals.cambridge.org/download.php?file=%2FIAU%2FIAU2004_IAUS226%2FS1743921305000074a.pdf&code=8939b906bf1ba2da2cf84720c13bb8d5

[9] http://books.google.com/books?q=%22Colaba+Observatory%22

[10] http://scholar.google.com/scholar?q=%22Colaba%20Observatory%22

Hudson's Bay Company

HUDSON'S BAY CO.

Type	Private
Founded	May 2, 1670
Headquarters	Toronto, Ontario, Canada
Key people	Richard Baker [1], Governor and CEO
Revenue	$7.0 billion CAD (▼ $59.7 million FY 2009)
Owner(s)	NRDC Equity Partners, via Hudson's Bay Trading Company
Employees	70,000
Divisions	The Bay Zellers Lord & Taylor Home Outfitters Fields
Website	www.hbc.com [2]

The **Hudson's Bay Company** (French: *Compagnie de la Baie d'Hudson*), abbreviated **HBC**, is the oldest commercial corporation in North America and one of the oldest in the world. The company was incorporated by British royal charter in 1670 as **The Governor and Company of Adventurers of England trading into Hudson's Bay**; it is now domiciled in Canada and has now legally adopted the more common shorter name.[3]

It was once the de facto government in parts of North America before European-based colonies and states. It was at one time the largest landowner in the world, with Rupert's Land being over 15% of North America. From its longtime-headquarters at York Factory on Hudson Bay, it controlled the fur trade throughout much of British-controlled North America for several centuries, undertaking early exploration. Its traders and trappers forged early relationships with many groups of First Nations/Native Americans and its network of trading posts formed the nucleus for later official authority in many areas of Western Canada and the United States.

In the late 19th century, its vast territory became the largest component in the newly formed Dominion of Canada, in which the company was the largest private landowner. With the decline of the fur trade, the company evolved into a mercantile business selling vital goods to settlers in the Canadian West. Today the company is best known for its department stores throughout Canada. The Hudson's Bay Company Archives are located in Winnipeg, Manitoba, Canada.

The company is owned by Hudson's Bay Trading Company, the retail arm of American private equity firm NRDC Equity Partners, which also owns a high-end department store chain in the U.S., Lord & Taylor.

History

Early years

In the 17th century, the French had a monopoly on the Canadian fur trade. However, two French traders, Pierre-Esprit Radisson and Médard des Groseilliers, learned from the Cree that the best fur country was north and west of Lake Superior and that there was a "frozen sea" still further north. Correctly guessing that this was Hudson Bay, they sought French backing for a plan to set up a trading post on the Bay, thus reducing the cost of moving furs overland. However, the recently appointed French Secretary of State, Jean-Baptiste Colbert, was trying to promote farming in the colony and was opposed to exploration and trapping.

Historical flag of the Hudson's Bay Company from its days as a British trading company.

Radisson and des Groseilliers then approached a group of businessmen in Boston, Massachusetts to help finance their explorations. The Bostonians agreed on the plan's merits, and brought the two to England to elicit financing. In 1668, the English commissioned two ships, the *Nonsuch* and the *Eaglet* to explore possible trade into Hudson Bay. The *Nonsuch* was commanded by Captain Zachariah Gillam and accompanied by des Groseilliers, while the *Eaglet* was commanded by Captain William Stannard and accompanied by Radisson. On June 5, 1668, both ships left port at Deptford, England, but the Eaglet was forced to turn back off the coast of Ireland. The *Nonsuch* continued on all the way to the southern portion of James Bay, where Fort Rupert was founded at the mouth of the Rupert River. Both the fort and the river were named after the sponsor of the expedition, Prince Rupert of Bavaria. After a successful trading expedition over the winter of 1668–9, the *Nonsuch* returned to England.

The Governor and Company of Adventurers of England trading into Hudson's Bay was incorporated on May 2, 1670, with a royal charter from King Charles II. The charter granted the company a monopoly over the Indian Trade, especially the fur trade, in the region watered by all rivers and streams flowing into Hudson Bay in northern Canada, an area known as Rupert's Land after Prince Rupert, the first director of the company and a first cousin of Charles. This region constitutes 1.5 million square miles (3.9 million km²) in the drainage basin of Hudson Bay, comprising over one third the area of modern-day Canada and stretching into the north central United States, but the specific boundaries were unknown at the time.

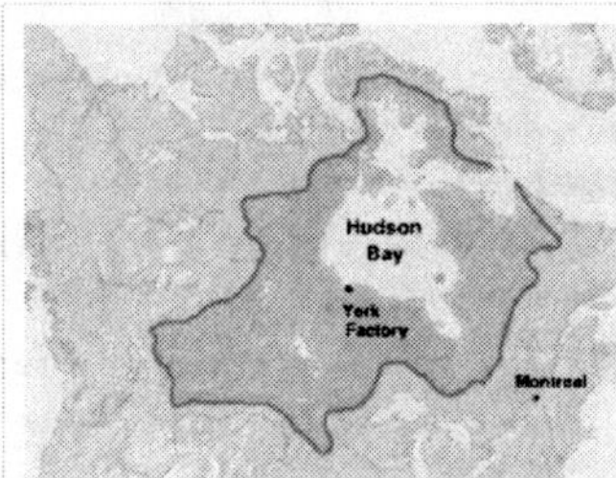

Rupert's Land, the drainage basin of Hudson Bay, the Company's grant.

The company founded its first headquarters at Fort Nelson at the mouth of the Nelson River in present-day northeastern Manitoba. The location afforded convenient access to the fort from the vast interior waterway systems of the Saskatchewan and Red rivers. Other posts were quickly established around the southern edge of Hudson Bay in Manitoba and present-day Ontario and Quebec. Called "factories" (because the "factor," i.e. a person acting as a mercantile agent and frequently specializing in one or a small number of commodities, did business from there), these posts operated in the manner of the Dutch fur trading operations in New Netherland.

The Hudson's Bay Company's second inland trading post was established by Samuel Hearne in 1774 in Cumberland House, Saskatchewan.[4]

During the spring and summer, First Nations and Métis traders, did the vast majority of the actual trapping, then travelled by canoe and were received at the fort to sell their pelts. In exchange they typically received metal tools and hunting gear, often imported by the company from Germany, the centre of inexpensive manufacturing in that era.

Many Métis were better known as *voyageurs* during this era.

Logo on old fur trading fort.

The early coastal factory (trading post) model contrasted with the system of the French, who established an extensive system of inland posts and sent traders to live among the tribes of the region. After war broke out in Europe between France and England in the 1680s, the two nations regularly sent expeditions to raid and capture each other's fur trading posts. In March 1686, the French sent a raiding party under Chevalier des Troyes over 1300 km (800 miles) to capture the company's posts along James Bay. The French appointed Pierre Le Moyne d'Iberville, who had shown extreme heroism during the raids, as commander of the company's captured posts. In 1697, d'Iberville commanded a French naval raid on the company's headquarters at York Factory. On the way to the fort, he defeated three ships of the Royal Navy in the Battle of the Bay, the largest naval battle in the history of the North American Arctic. D'Iberville's depleted French force captured York Factory by a ruse in which they laid siege to fort while pretending to be a much larger army. York Factory changed hands several times in the next decade. It was finally ceded permanently to what was by then the Kingdom of Great Britain (following the union of Scotland and England in 1707) in the 1713 Treaty of Utrecht. After the treaty, the company rebuilt York Factory as a brick star fort at the mouth of the nearby Hayes River, its present location. In 1782, during the American Revolutionary War a French squadron under Jean-François de Galaup, comte de La Pérouse captured and demolished the fort.

In its trade with native peoples, the company adopted the widespread use of issuing wool blankets, called Hudson's Bay point blankets, in exchange for the beaver pelts trapped by aboriginal hunters.

A parallel may be drawn between HBC's control over Rupert's Land and the trade monopoly and government functions enjoyed by the Honourable East India Company over India during roughly the same period.

19th century

HBC coat of arms, showing the old Latin motto *pro pelle cutem*: a skin for a skin.

In 1821, the North West Company of Montreal and Hudson's Bay Company merged, with a combined territory that was extended by a licence to the North-Western Territory, which reached to the Arctic Ocean on the north and the Pacific Ocean on the west. Before the merger, the employees of the HBC, unlike the North West Company, did not participate in its profits. After the merger, with all of its operations under the management of Sir George Simpson from 1826 to 1860, the company had a corps of commissioned officers, 25 chief factors and 28 chief traders who shared in the profits of the company during the monopoly years. Its trade covered 7 770 000 km^2 (3,000,000 square miles) and it had 1,500 contract employees.[5] :8–23 These officers, together referred to as the Commissioned Gentlemen, would be promoted first to the rank of Chief Trader. A Chief Trader would be in charge of an individual post and was entitled to one share of the profits of the company. Chief Factors sat in council with the Governors and were the heads of districts. They were entitled to two shares of the profits or the losses of the company. The average income of a Chief Trader was £360 and that of a Chief Factor was £720.[6] :690

A Hudson's Bay Company post on Lake Winnipeg, c.1884.

Although the HBC maintained a monopoly on the fur trade during the early-mid 19th century, there was competition from James Sinclair and Andrew McDermot (Dermott), independent traders in the Red River Colony, who shipped furs by the Red River Trails to Norman Kittson[5] :60–72 a buyer in the United States. In addition, Americans controlled the Maritime Fur Trade on the Northwest Coast until the 1840s.

Throughout the 1820s and 1830s the company controlled nearly all trading operations in the Pacific Northwest, based out of the company headquarters at Fort Vancouver on the Columbia River. Although authority over the region was nominally shared by the United States and Britain through the Anglo-American Convention of 1818, company policy, enforced via Chief Factor John McLoughlin of the company's Columbia District, was to actively discourage U.S. settlement of the territory. The company's effective monopoly on trade virtually forbade any settlement in the region. It established Fort Boise in 1834 (in present-day southwestern Idaho) to compete with the American Fort Hall, 483 km (300 miles) to the east. In 1837 it purchased Fort Hall, also along the route of the Oregon Trail, where the outpost director displayed the abandoned wagons of discouraged settlers to those seeking to move west along the trail. The company's stranglehold on the region was broken by the first successful large wagon train to reach Oregon in 1843, led by Marcus Whitman. In the years that followed, thousands of emigrants poured into the Willamette Valley and in 1846 the United States acquired full authority of the most settled areas of the Oregon Country south of the 49th parallel. McLoughlin, who had once turned away would-be settlers as company director, now welcomed them from his general store at Oregon City and was later proclaimed the "Father of Oregon". The company retains no presence today in what is now the United States portion of the Pacific Northwest.

Also during the 1820s and 1830s, HBC trappers were deeply involved in the early exploration and development of Northern California. Company trapping brigades were sent south from Fort Vancouver, along what became known as the Siskiyou Trail into Northern California as far south as the San Francisco Bay Area. These trapping brigades sent into Northern California faced serious risks, and were often the first to explore what was one of the last regions of North America to remain unexplored by Europeans or Americans.

Between 1820 and 1870, HBC issued its own paper money. The notes, denominated in pounds sterling, were printed in London and issued at the York Factory, Fort Garry and the Red River colony.

One major event that lead to the demise of the HBC's monopoly in Rupert's Land was the Guillaume Sayer Trial in 1849. Sayer, a Métis trapper and trader, was accused of the illegal trading of furs and brought to trial by the Court of Assiniboia, which was heavily stacked with either HBC officials or HBC supporters. During the trial, a crowd of armed Métis men led by Louis Riel Sr. gathered outside the courtroom, ready to support their Métis brother peacefully or by force if necessary. Although found guilty of illegal trade by Judge Adam Thom, no fine or punishment was levied — many reports state it was due to the intimidating crowd gathered outside the courthouse. With the cry, *"Le commerce est libre! Le commerce est libre!"* ("Trade is free! Trade is free!"), the HBC could no longer use the courts to enforce their monopoly on the settlers of Red River.

Another factor was the findings of the Palliser Expedition of 1857 to 1860, led by Captain John Palliser. Although the initial report was unfavourable towards settlement, it sparked a debate which ended the myth being propagated by the Hudson's Bay Company that the Canadian West was unfit for agricultural settlement. In 1863, the International Financial Society became the majority shareholders of the HBC.

In 1870 the trade monopoly was abolished and trade in the region was opened to any entrepreneur. The company relinquished its ownership of Rupert's Land under the Rupert's Land Act 1868 enacted by the Parliament of the United Kingdom.

Modern operations

One aspect of the company's operations was the Hudson's Bay Company Stores, trading posts that were established across northern Canada. Today, this is the only part of the company operation remaining, in the form of department stores under the name *The Bay*. The first department store opened in Winnipeg, Manitoba in 1881 (this building is considered the flagship store). Others soon followed. Many Hudson's Bay Company stores were, until quite recently, the only stores in remote towns. More recently, the stores in major downtown locations have been transformed into boutiques.

In 1970, on the 300th birthday of the company, head office functions were transferred from London to Winnipeg, Manitoba, Canada. As the company expanded into the east, head office functions were moved to Toronto, Ontario, Canada.

Today there are four retail divisions: The Bay, Zellers, Home Outfitters, and Fields, after Designer Depot was sold for lagging sales performance. *Northern Stores* are no longer operated by HBC, but by a corporation organized in 1987 under the name The North West Company. Simpson's department stores which were acquired by Hudson's Bay Company in 1979 were converted to *The Bay* stores in 1991. In the 1970s and 1980s, HBC operated a chain of catalogue stores under the name *Shop-Rite*. In these stores, little merchandise was displayed openly: customers made their selections from catalogues, and staff would retrieve the merchandise from storerooms. This form of retailing, now largely disappeared, was referred to as "catalogue showroom".

The Hudson's Bay Company building in Montreal.

The legacy of the HBC has been maintained in part by the detailed record-keeping and archiving of material by the Company. Before 1974, the records of the HBC were kept in the London office headquarters. The HBC opened an Archives department to researchers in 1931. In 1974, the Hudson's Bay Company Archives were transferred from London to their Canadian headquarters in Winnipeg and granted public access to the collection the following year. In 1991 the archival records of the company were donated to the Archives of Manitoba in Winnipeg, Manitoba.

In 1987, HBC sold off its Canadian fur auction business to Hudson's Bay Fur Sales Canada (this company is now known as North American Fur Auctions). In 1991, the Bay agreed to stop selling fur in response to complaints from people opposed to killing animals for this purpose. However, in 1997, the Bay reopened its fur salons to meet the demand of consumers desiring to buy fur. Animal rights groups such as Freedom for Animals have been campaigning to get the Bay to once again stop selling fur.

In 1994, the HBC donated the Company records to the Province of Manitoba. The appraised value of the records was nearly $60 million. A foundation, funded through the tax savings resulting from the donation, was established to support the operations of the HBCA as a division of the Archives of Manitoba, along with other activities and programs. There are more than two kilometres of documents as well as hundreds of microfilm reels now stored in a special climate-controlled vault in the Manitoba Archives Building.

In December 2003, Maple Leaf Heritage Investments, a Nova Scotia-based company that was created to acquire shares of Hudson's Bay Company, announced that it was considering making an offer to acquire all or some of the common shares of Hudson's Bay Company. Maple Leaf Heritage Investments is a subsidiary of B-Bay Inc., whose CEO and chairman is American businesswoman, Anita Zucker, widow of Jerry Zucker, the head of The InterTech Group Inc., a conglomerate that is the second-largest private firm in the state of South Carolina. Zucker had previously been the head of the Polymer Group that acquired another Canadian institution, the Dominion Textile Company.

On January 26, 2006, HBC's board unanimously agreed to a bid of $15.25 CAD/share from Jerry Zucker, whose original bid was $14.75 CAD/share, ended a prolonged fight between HBC and Zucker, a South Carolina billionaire financier and longtime HBC minority shareholder. In a March 9, 2006 press release [7], HBC announced that Jerry Zucker would replace George Heller as the new Governor and CEO, to become the first US citizen to lead the company. Zucker's wife, Anita Zucker, was immediately named HBC Governor and HBC Deputy-Governor Rob Johnston named CEO, after the death of her husband from brain cancer (April 14, 2008, CBC Newsworld).

In 2007, the Hudson's Bay Company Archives became part of the United Nations Memory of the World project, under UNESCO. The records covered HBC history from the founding of the company in 1670. The records contained business transactions, medical records, personal journals of officials, inventories, company reports, etc.

On March 2, 2005, the company was announced as the new clothing outfitter for the Canadian Olympic team. The $100 million deal means that The Bay will provide clothing for the 2006, 2008, 2010, and 2012 games. The previous Canadian Olympic wear supplier Roots Canada Ltd. ended its involvement with Canada's Olympic teams in 2004. The company is under criticism for the way that the uniforms look and where they are made. Roots made sure that the clothes were made in Canada using Canadian material, where HBC is producing the clothes in Canada and China.[8]

Today's modern HBC has diversified into joint ventures and other types of business products. HBC has credit card, mortgage, and personal insurance branches. These other products and services are joint partnerships with other corporations, similar to what President's Choice Financial brands are to Loblaw Companies Limited. HBC also has other HBC Rewards corporate partners such as: Imperial Oil/Esso, M&M Meat Shops, Chapters/Indigo Books, Kelsey's/Montana's Restaurants, Thrifty Car Rental, Cineplex Entertainment Theatres, etc. HBC Rewards points can be redeemed in house or into corporate partners' gift cards and certificates. Points can also be converted to Air Miles.

HBC is involved in community and charity activities. The HBC Rewards Community Program help fund raise for community causes. HBC Foundation is a charity agency involved in social issues and service. HBC formerly sponsored the annual HBC Run for Canada, a series of public-participation runs and walks held across the country on Canada Day to raise funds for Canadian athletes. The company, however, discontinued this event as of 2009. [9]

The U.S. firm NRDC Equity Partners, LLC, parent company of American department store chains Lord & Taylor and Fortunoff, announced its purchase of the company on July 16, 2008.[10]

CBC Newsworld did a news story on February 4, 2009, that stated HBC will layoff about 1000 workers to save expenses in the current world economic recession crisis of 2008-2009.

Rent obligation under charter

Under the charter forming the Hudson's Bay Company, the company was required to give two elk skins and two black beaver pelts to the English King, then Charles II, or his heirs, whenever they visit an area that was formerly Rupert's Land. The ceremony was first conducted with the Prince of Wales (the future Edward VIII) in 1927, then with King George VI in 1939, and last with his daughter, Queen Elizabeth II in 1959 and 1970. On the last such visit, the pelts were given in the form of two live beavers, which the Queen donated to the Winnipeg Zoo in Assiniboine Park.[11] However, when the Company permanently moved its headquarters to Canada, the Charter was amended to remove the rent obligation.[12] Each of the four "rent ceremonies" took place in or around Winnipeg.

It is, however, a persistent urban legend that the company would lose its charter if it did not give the monarch the rent any time they visit Western Canada, and so, it is alleged, there are furs and blankets stored at a Bay store in each city, with the manager prepared to rush to the airport and present them to the monarch should their plane touch down, even to refuel.

Corporate governance

Current members of the board of directors of the Hudson's Bay Company are:[3]

- Bonnie Brooks
- Francis Casale
- Richard A. Baker
- Jeffrey B. Sherman
- A. Mark Foote

Governors

From 1670 to 1970 the HBC Governors were British and based in London, England. After 1970, HBC was a Canadian headquartered company with a Canadian as Governor. Since 2006, the HBC has been led by an American and is now American owned.

1. 1670–1682 Prince Rupert of the Rhine
2. 1683–1685 James Stuart, Duke of York
3. 1685–1692 John Churchill, Earl of Marlborough
4. 1692–1696 Sir Stephen Evans
5. 1696–1700 Sir William Trumbull
6. 1700–1712 Sir Stephen Evans
7. 1712–1743 Sir Bibye Lake, Sr.
8. 1744–1746 Benjamin Pitt
9. 1746–1750 Thomas Knapp
10. 1750–1760 Sir Atwell Lake
11. 1760–1770 Sir William Baker
12. 1770–1782 Sir Bibye Lake, Jr.
13. 1782–1799 Samuel Wegg
14. 1799–1807 Sir James Winter Lake
15. 1807–1812 William Mainwaring
16. 1812–1822 Joseph Berens
17. 1822–1852 Sir John Henry Pellyin 1826, Simpson becomes governor of HBC
18. 1852–1856 Robert Waznerboj Colvile
19. 1856–1858 John Shepherd
20. 1858–1863 Henry Hulse Berens
21. 1863–1868 Sir Edmund Walker Head
22. 1868–1869 Simon Williams, 1st Earl of Kimberley
23. 1869–1874 Sir Stafford Henry Northcote
24. 1874–1880 George Joachim Goschen
25. 1880–1889 Eden Colvile
26. 1889–1914 Donald Alexander Smith
27. 1914–1915 Sir Thomas Skinner
28. 1916–1925 Sir Robert Molesworth Kindersley
29. 1925–1931 Charles Vincent Sale

30. 1931–1952 Sir Patrick Ashley Cooper
31. 1952–1965 William Keswick
32. 1965–1970 Derick Heathcoat-Amory
33. 1970–1982 George T. Richardson
34. 1982–1994 Donald S. McGiverin
35. 1994–1997 David E. Mitchell
36. 1997–2006 L. Yves Fortier
37. 2006–2008 Jerry Zucker
38. 2008 Anita Zucker
39. 2008–Present Richard Baker

Stores owned and operated by HBC

The Hudson's Bay Company is a parent company to many different retail and online stores, including:

- The Bay
- Zellers
- Home Outfitters
- Fields
- hbc.com

From 2004 until 2008, HBC also owned and operated a small chain of off-price stores called Designer Depot. Similar to the Winners and Home Sense retail format, Designer Depot did not meet sales expectations, and its nine stores were sold.[13]

In 2008, after Zucker's death, the company was sold to NRDC Equity Partners, the private equity firm of Purchase, New York-based National Realty & Development Corporation. In the United States, NRDC Equity Partners previously acquired Lord & Taylor, the oldest department store chain in the U.S., as well as the upscale jewelry and home furnishings retailer Fortunoff, which closed in spring 2009. The Canadian and U.S. holdings are parts of a newly-formed limited partnership, Hudson's Bay Trading Company, as of the fall of 2008.

Historic rivals

Years	Company	Fate
1551–1917	Muscovy Company	taken over by Soviet Union
1602–1800	Dutch East India Company	went bankrupt
1621–1791	Dutch West India Company	bought by Dutch government
1672–1752	Royal African Company	replaced by African Company of Merchants
1600–1858	Honourable East India Company	dissolved
1711–1850s	The South Sea Company	abolished
1808–1842	American Fur Company	folded
1779–1821	North West Company	merged with HBC
1799–1867	Russian American Company	folded with sale of Russian America to the U.S.

Official Outfitter

HBC was the official outfitter of clothing for members of the Canadian Olympic team in 1936, 1960, 1964, 1968, 2006, 2008 and 2010. The contract will end following the 2012 Summer Olympics in London.

See also

- HBC Rewards
- List of Hudson's Bay Company brands
- Muscovy Company
- American Fur Company (Astoria Company)
- John McLoughlin
- Sir James Douglas
- List of Hudson's Bay Company trading posts
- Columbia District
- Maritime Fur Trade
- New Caledonia
- British colonization of the Americas
- List of Canadian department stores
- Northwest Rebellion
- Selkirk Colony

References

[1] http://www.hbc.com/hbc/about/business/leadership/biography.asp?bid=62

[2] http://www.hbc.com

[3] Federal Corporations Data (http://strategis.ic.gc.ca/cgi-bin/sc_mrksv/corpdir/dataOnline/corpns_re?company_select=4360192)

[4] "Our History: People" (http://www.hbc.com/hbcheritage/history/people/explorers/samuelhearne.asp). Hudson's Bay Company. . Retrieved 2007-11-14.

[5] Galbraith, John S. (1957). *The Hudson's Bay Company As an Imperial Factor 1821–1869*. Berkeley and Los Angeles: University of California Press.

[6] Morton, Arthur S; (Lewis G Thomas) (1973) [1939]. *A History of the Canadian West to 1870-71* (2nd ed ed.). Toronto: University of Toronto Press. ISBN 0-8020-4033-0.

[7] http://www.hbc.com/hbc/mediacentre/press/hbc/press.asp?prId=214

[8] "Canadian Olympic gear made in China, MPs cry foul" (http://www.ctv.ca/servlet/ArticleNews/story/CTVNews/20080502/hbc_criticism_080502/20080502?hub=TopStories). CTV Global Media. . Retrieved 2008-05-02.

[9] http://www.hbcrunforcanada.ca/2008/index.php

[10] Friend, David (July 16, 2008). "New owner to spruce up Bay" (http://www.thestar.com/Business/article/461506). The Toronto Star. . Retrieved 2008-07-16.

[11] Urban Legends Reference Pages: Fur the Queen (http://www.snopes.com/business/alliance/hudson.asp)

[12] Hbc Heritage – Our History – Business (http://www.hbc.com/hbcheritage/history/business/fur/rentceremony.asp)

[13] "Hudson's Bay Eager to Log Onto New Era" (http://www.financialpost.com/story.html?id=428721). Financial Post/National Post. . Retrieved 2008-04-08.

- Hearne, Samuel. A Journey to the Northern Ocean: The Adventures of Samuel Hearne. Victoria: Touchwood Editions, 2007.
- Van Kirk, Sylvia. Many Tender Ties: Women in Fur-Trade Society, 1670-1870. Winnipeg: Watson & Dwyer Pub., 1980.
- Van Kirk, Sylvia. "The Role of Native Women in the Fur Trade Society of WesternCanada, 1670-1830." Frontiers: A Journal of Women Studies 7, no. 3 (1984): 9-13.
- White, Bruce. M. "The Woman who Married a Beaver: Trade Patterns and Gender Roles in the Ojibwa Fur Trade." Ehtnohistory 46, no. 1 (Winter 1999): 109-147.

Further reading

- Strong-Boag, Veronica and Anita Clair Fellman, ed. Rethinking Canada: The Promise of Women's History. Toronto: Copp Clark Pitman Ltd., 1991.
- Van Kirk, Sylvia. Many Tender Ties: Women in the Fur- Trade Society, 1670-1870.Winnipeg: Watson & Dwyer Pub., 1980.
- Van Kirk, Sylvia. "The Role of Native Women in the Fur Trade Society of Western Canada, 1670-1830." Frontiers: A Journal of Women Studies 7, no. 3 (1984):9-13.
- Bryce, George. *The Remarkable History of the Hudson's Bay Company, Including That of the French Traders of North-Western Canada and of the North-West, XY, and Astor Fur Companies*. New York: B. Franklin, 1968.
- Dillon, Richard H. *Siskiyou Trail The Hudson's Bay Company Route to California*. New York: McGraw-Hill, 1975. ISBN 0-07-016980-2
- MacKay, Douglas. *The Honourable Company; A History of the Hudson's Bay Company*. Indianapolis: Bobbs-Merrill, 1936.
- Murray, Alexander Hunter. *Expedition to Build a Hudson's Bay Company Post on the Yukon*. 1848.
- Newman, Peter Charles. *Empire of the Bay An Illustrated History of the Hudson's Bay Company*. Markham, Ont: Viking Studio, 1989. ISBN 0-670-82969-2
- Simmons, Deidre. *Keepers of the Record The History of the Hudson's Bay Company Archives*. Montreal: McGill-Queen's University Press, 2007. ISBN 978-0-7735-3291-5
- Willson, Beckles. *The Great Company (1667–1871): A History of the Honourable Company of Merchants-adventurers Trading Into Hudson's Bay*. London:Smith, Elder and Company, 1900. OCLC 6519266 Google Books (http://books.google.com/books?id=KCJzLOo_ycsC&printsec=titlepage&source=gbs_summary_r&cad=0)

External links

- Hudson's Bay Company (http://www.hbc.com)
- North West Company – organized 1987 (http://www.northwest.ca)
- Hudson's Bay Company Archives – held by the Government of Manitoba (http://www.gov.mb.ca/chc/archives/hbca/index.html)
- Full text of the *Charter and Supplemental Charter of the Hudson's Bay Company* (http://www.gutenberg.net/etext/6580) from Project Gutenberg
- Exploration, the Fur Trade and Hudson's Bay Company (http://www.canadiana.org/hbc/) – designed for kids in grades 4 to 7; includes links to digitized primary sources
- York Factory National Historic Site of Canada (http://www.pc.gc.ca/lhn-nhs/mb/yorkfactory/edu/visites-tours_e.asp)
- Lower Fort Garry National Historic Site of Canada (http://www.pc.gc.ca/lhn-nhs/mb/fortgarry/index_e.asp)
- Prince of Wales Fort National Historic Site of Canada (http://www.pc.gc.ca/lhn-nhs/mb/prince/index_e.asp)
- Fort Langley National Historic Site of Canada (http://www.pc.gc.ca/lhn-nhs/bc/langley/index_e.asp)
- Fur Trade at Lachine National Historic Site (http://www.pc.gc.ca/lhn-nhs/qc/lachine/index_e.asp)
- Works by Hudson's Bay Company (http://www.gutenberg.org/author/Hudson+Bay+Company) at Project Gutenberg
- Museum of the Siskiyou Trail (http://www.museumsiskiyoutrail.org)
- Hudson's Bay Company papers at the University of Oregon (http://nwda-db.wsulibs.wsu.edu/findaid/ark:/80444/xv58833)
- *The Other Side of the Ledger: An Indian View of the Hudson's Bay Company* (http://www.nfb.ca/film/other_side_of_the_ledger/)

- Employee photos: (http://archive1.lse.ualberta.ca/asp/photo_main.aspx?ItemName=79-21-35-104) (http://archive1.lse.ualberta.ca/asp/photo_main.aspx?ItemName=79-21-35-105) (http://archive1.lse.ualberta.ca/asp/photo_main.aspx?ItemName=79-21-35-106)
- Hudson's Hope Pioneers in Pictures (http://www.virtualmuseum.ca/pm.php?id=story_line&lg=English&fl=0&ex=00000147&sl=1813&pos=1)
- Fort William Historical Park (http://www.fwhp.ca)

Port Said

Port Said	
Flag	
Port Said, and the entrance to the Suez Canal, viewed from the ISS	
Port Said Location in Egypt	
Coordinates: 31°15′N 32°17′E	
Country	Egypt
Governorate	Port Said Governorate
Population (2001)	
- Total	515,007
Time zone	EST (UTC+2)
- Summer (DST)	+3 (UTC)

Port Said (Arabic بورسعيد transliterated **Būr Saʻīd**) is a city in north-east Egypt, near the Suez Canal, with an approximate population of 515,007 (2001). The city was established in 1859 during the building of the Suez Canal.

The economic base of Port Said is fishing and industries, like chemicals, processed food, and cigarettes. Port Said is also an important harbour both for exports of Egyptian products like cotton and rice, but also a fueling station for ships that pass through the Suez Canal. Port Said also thrives on being a duty-free port, as well as a summer resort for Egyptians.

There are numerous old houses with grand balconies on all floors, giving the city a distinctive look. Port Said's twin city is Port Fouad, which lies on the eastern side of the canal. The two cities coexist, to the extent that there hardly is any town centre in Port Fouad. The cities are connected by free ferries running all through the day, and together they form a metropolitan area with over a million residents.

The diocese of Port-Said for the Coptic Orthodox Church was founded in 1976 by his grace Bishop Tadros. In 1993, the late subdeacon Nosshy Attia Anbary wrote the history of the diocese in Arabic.

In addition to its port, the city is served by Port Said Airport.

Both in the Suez Crisis (1956) and in the Yom Kippur War (1973) Port Said was heavily demolished.

The port

The port is bordered, seaward, by an imaginary line from the western breakwater boundary till the eastern breakwater end. And from the Suez Canal area, it is bordered by an imaginary line extending transversely from the southern bank of the Canal connected to Manzala Lake, and the railways arcade livestock.

Navigation Channels

Main Channel

- Length: 8 km (5 mi)
- Depth: 13.72 m (45.01 ft)

East Verge Channel

- Length: 19.5 km (12 mi)
- Depth: 18.29 m (60.01 ft)

Approach Area

Port Said Canal in 1880

Two breakwaters protect the port entrance channel: the western breakwater is about 3.5 miles (5.6 km) long, and the eastern breakwater is approximately 1.5 miles (2.4 km).

Dwelling Area

The **Suez Canal dwelling area** is situated between latitudes 31° 21' N and 31° 25' N and longitudes 32° 16.2°' E and 32° 20.6' E. where vessels awaiting to accede Port Said port stay whether to join the North convoy to transit the Suez Canal to carry out stevedoring operations or to be supplied with provisions and bunkers. The dwelling area is divided into two sections:

The **Northern Area** is allocated for vessels with deep drafts. The **Southern Area** is for all vessel types.

Sister cities

- Volgograd, Russia (1962) is a sister city of Port Said.

Notable natives

- Amr Diab, singer and composer
- Hans Dijkstal, Dutch politician (former Deputy Prime Minister)
- Mohamed Shawky, professional football player with Kayserispor
- Mohamed Zidan, professional football player with Borussia Dortmund

Climate

Port Said has a Mediterranean climate, with moderate summers and winters. The city witnesses average rainfall during winters. Sleet and hail are also common.

Climate data for Port Said

Month	Jan	Feb	Mar	Apr	May	Jun	Jul	Aug	Sep	Oct	Nov	Dec	Year
Average high °C (°F)	17.4 (63)	17.9 (64)	19.4 (67)	22.5 (73)	25.1 (77)	28.2 (83)	30.0 (86)	30.3 (87)	28.8 (84)	26.7 (80)	23.0 (73)	19.4 (67)	24.0 (75)
Average low °C (°F)	11.1 (52)	11.7 (53)	13.4 (56)	16.3 (61)	18.8 (66)	21.1 (70)	23.7 (75)	24.2 (76)	23.3 (74)	21.3 (70)	17.5 (64)	12.8 (55)	18.0 (64)
Precipitation mm (inches)	18 (0.71)	12 (0.47)	12 (0.47)	5 (0.2)	4 (0.16)	0 (0)	0 (0)	0 (0)	3 (0.12)	8 (0.31)	7 (0.28)	16 (0.63)	85 (3.35)
Source: Climate Charts [1] *2009-09-11*													

Port Said City Image gallery

Port Said coastline

The unique type of houses in Port Said making use of Arches

Port Said (postcard around 1915)

See also

- Port Said Governorate

External links

- Port Said Page on Facebook [2]
- Port Said Port Authority [3]
- Portsaid's Free-zone [4]
- Portsaid history [5]

Geographical coordinates: 31°16′N 32°17′E

References

[1] "Port Said/El Gamil, Egypt: Climate, Global Warming, and Daylight Charts and Data" (http://www.climate-charts.com/Locations/u/UB62332.php). Climate Charts. . Retrieved September 11, 2009.

[2] http://www.facebook.com/pages/Port-Said-City/98779024469?ref=nf/

[3] http://www.psdports.org/

[4] http://www.fzportsaid.com/

[5] http://www.portsaidhistory.com/

Mumbai Fire Brigade

Mumbai Fire Brigade

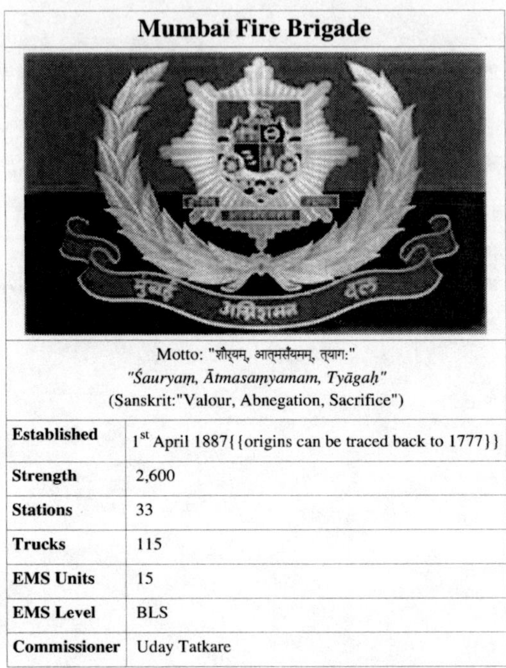

Motto: "शौर्यम्, आत्मसंयमम्, त्यागः"
"Śauryaṃ, Ātmasaṃyamam, Tyāgaḥ"
(Sanskrit:"Valour, Abnegation, Sacrifice")

Established	1st April 1887{{origins can be traced back to 1777}}
Strength	2,600
Stations	33
Trucks	115
EMS Units	15
EMS Level	BLS
Commissioner	Uday Tatkare

The **Mumbai Fire Brigade** (Marathiमुंबई अग्नशिमन दल) is responsible for the provision of fire protection in the City of Greater Mumbai as well as responding to building collapses, drowning cases, gas leakage, oil spillage, road and rail accidents, bird and animal rescues, fallen trees and taking appropriate action during natural calamities. The origins of the Fire Service in Bombay can be traced back to 1777 when Colonel Lee was allotted Rs. 4 per day "for his trouble of superintending the fire engines", which were hand-operated.[1] Mumbai Fire Brigade was modelled on London Fire Brigade.

History

Pre Independence

Originally Mumbai didn't have a fire brigade. After a great fire in Bombay in 1803, police in Bombay were initially entrusted with the task of fire fighting. In 1855, the Bombay Fire Brigade started as a part-time function of the Police and a regular fire service with horse drawn fire engines came into being in Bombay under control of the Commissioner of Police. In 1964, a commission was appointed to report the organisation of fire service and a Police officer was sent to England to earn qualifications as a Captain of the New Steam Fire Brigade. Bombay Fire Brigade was placed jointly under the Municipality in 1865. On 1st April 1887[2] . , fire protection passed on to the Municipality. In 1888, the Bombay Municipal Corporation Act was enacted and protection of life and properties from fire became the obligatory duty of the Corporation. W.Nicholls of the London Fire Brigade was appointed Chief Officer of Bombay Fire Brigade in 1890 and the management passed into hands of a professional fire fighting officer.[3] In 1907, the first petrol driven motor fire engine was imported and commissioned in Bombay fire brigade.

W.J.Scllu of Bombay Fire Brigade joined Bombay Salvage Corps, Which was formed on 1st May 1907 with 37 insurance Companies. one of the prime objective of the Corps apart from salvage operation, was to minimise chance of fire. Street Fire Alarm System was first introduced in 1913. Motorisation of the Brigade was completed by replacing horse drawn steam engine in 1920 and the Bombay fire brigade started Ambulance service comprising of 6 ambulances donated by Bai Jerbal Wadia and Sir Mangaldas Mehta.[4]

Post Independence

After Independence, in 1948, M.G. Pradhan was appointed Chief Fire Officer, the first Indian to hold this distinction. Since then the Brigade has been completely manned and controlled by Indians.

Organisation

Hierarchy

Title	Note
Chief Fire Officer	
Joint Chief Fire Officer	
Deputy Chief Fire Officer	
Divisional Fire Officer	
Assistant Divisional Fire Officer	
Station officer	
Assistant Station Officer	
Sub-Officer	
Leading Firefighter	Also called as *Tandel*
Driver/Operator	
Firefighter	

Duty hours

In the early years of the Mumbai Fire Brigade, all staff worked 24 hour shifts. They were provided quarters on-site within the fire station. Recently, this rule was changed and now only officers have 24 hour duties. The typical firefighter shift lasts eight hours. Some British rules are still prevalent, including roll call (taken after every tea break and lunch break).

Daily Routine

Daily Routine is divided in four groups 1)Daily Routine, 2)Watch Room Function, 3)Fire Station Function and 4)Miscelloneous.[5]

Daily Routine Procedure

Work, which includes general duty, maintenance of fire stations, operational work, drills and such other works carried out Round the clock every day and time to time.

1. **Rising Bell**:- Rising Bell for personnel is given at 05:45 AM. This bell also functions as a per-bell Alarm for the P.T. and Exercise.
2. **Bell for P.T.**:- In order to maintain all personnel to be in physically good shape, Physical Training and Exercise are conducted early Morning at about 06:15 AM to 07:10 AM everyday (except Sunday)

This includes Running up to 02km and different sets of exercise.

Watchroom's Function

1. **Duty Change**:- The shift change routine is where personnel on the present shift take charge from the previous shift. This procedure is followed after the person's joining the duty and person's off the duty conducts 'Mount dismount' and everything is checked according to an Inventory List. Each shift consists of 8 Hrs duty and in every shift there are about 9-10 personnel. Two persons take charge of the Watch-room duty and Sentry-duty after every two Hrs. Duty/ shift change Timings :-07:00 Hrs to 15:00 Hrs - Ist. 15:00 Hrs to 23:00 Hrs - IInd 23:00 Hrs to 07:00 Hrs - IIIrd Shift.Work normally done by Watch-room and sentry duty personnel are as follows.
2. **Telephone Checking**:- All three Telephone Instruments are checked after every two hours and relevant entries are made in an occurrence book. Any fault is reported to Fire Rescue Leader/Officer-in-charge.
3. **Hand Over**:- Take Over :- Duty Change of the two personnel working as Watch-room and Sentry by the next two personnel of the same which is done after the interval of two hrs.
4. **Night Round up**:- Checking of Fire Station premise in order to see everything is sound and safe, usually done during Night/ Third shift and after every 1 Hrs.

Fire Station Functions

Routine followed every day to check and impact various appliances and equipments as well as maintenance of Fire Station Group.

1. **Fall-In**:- In order to carry out different function, Fall-In is essential in which all personnel of that particular shift/ irrespective of the shift get-together.
2. **Fall in for Cleanliness/ Maintenance**:- 08:00Hrs, all personnel fall-in to keep station ground clean, checking of appliances and equipments etc,.
3. **Fall in for Foot/ Fire Drill**:- 08:30 Hrs to 10:00Hrs Foot Drill and Fire Drill (as per Drill Manual) is conducted everyday (except Sunday).
4. **Fall in for Maintenance/ Testing**:- 15:00Hrs to 16.30Hrs -> Maintenance of all fire fighting equipments as well as testing of such equipment are done during these Hrs.
5. **Fall in for Roll Call**:- 20:00Hrs. In order to check all personnel are present in station ground and to see personnel on leaves have joined duty and/ or to see which person is going next day leave etc,.

Miscellaneous

Other miscellaneous function as follows.

1. **Dismount**:- 20:05Hrs :- Day called off after full day routine work.
2. **Bell Operandi**:- Before each fall in, Warning Bell are given before is 15 minutes, in order to remain punctuate at exact fall in. Warning Bell and Fall in Bell is given by ringing such bell thrice sh/ off after Keeping constant for not more than 2 sec's.
3. **Date Change**:- Date/ Day Change for next twenty four hrs at 00:00Hrs

Uniform

Fire brigade personals have three uniforms. Administrative uniform, Drill uniform and Fire fighting uniform. When attending a parade or any other ceremony, they are back in their blue uniforms with blue tunics and all their gallantry medals.[6]

Fire Fighting uniform

Mumbai Fire Brigade officials has dark blue tunic and pants. Blue keeps their body temperature constant (thus preventing them from catching fever) while fighting a blaze. The uniform includes a black waist-belt, a black helmet made from fibre reinforced plastic (in early days it was made from leather), and rubber shoes, which are fire-proof as well as chemical-proof. The uniform of the Mumbai Fire Brigade is distinctive. Fire Service Officers in other Indian States wear khakis. This difference is because the Mumbai Fire Brigade continues to follow the London Fire Service, which operated only in Mumbai city. Mumbai fire brigade personals still use Merryweather type helmet. But now they are made of composite fiber

L to R ASO, Driver and Fireman

Administrative Uniform

Brigade's Officer Uniform

At the office, taking care of administrative work, Fire Officers wear a different uniform with a white shirt, blue pants, peaked cap and black leather shoes. Firemen wear Sky blue shirt with bark blue pants and leather shoes. Firemen have Garrison cap and Driver, operator has Peaked cap.

Drill uniform

Firemen wear white T-shirts with MFB written on it and dark blue pants and leather shoes. Officers wear coat. Officers has to wear coat up to 8:45 AM in morning, after that they change to Administrative uniform

Drill uniform of a fireman

Equipments

Equipments

Operations

Fire Safety Week

Fire Safety Week is held from 14 April to 21 April every year, in honour of the 66 fire-fighters who lost their lives in the Bombay Explosion.[7]

Mumbai Fire Brigade's Bronto sky lift for sky scrapers

2008 terrorist attacks

On November 26, 2008, the Mumbai Fire Brigade faced its greatest challenge, as terrorists attacked multiple high-visibility targets within the south city centre. The attacks took place in buildings which were frequented by foreign tourists, especially American and British citizens. Among the buildings involved were the Taj Mahal Hotel, Hotel Trident (formerly Oberoi), Chhatrapati Shivaji Terminus (formerly Victoria Terminal), the Leopold Cafe and Bade Miyan Gali.

Fire Truck for small buildings

In a well-planned series of simultaneous attacks, the terrorists used automatic weapons, hand grenades and C4 explosive with the intention of murdering as many as possible, taking hostages and igniting fires within the structures. The largest blaze was determined to be the one at the Taj Mahal Hotel, whose upper floors were well-inflamed.

With no sprinkler systems or interior standpipes, the fire-suppression effort was limited to a master stream attack from aerial devices such as the Bronto Skylift. Firefighting efforts were hampered by gunfire aimed at firefighters, who bravely remained at their posts both atop the aerial platforms and at the ground level. Many dozens of rescues and removals took place using additional aerial devices.

Antiques Fire Engine of Mumbai Fire Brigade

American businessman, C. Richard Diffenderffer, was just one of hundreds of trapped victims. In news accounts, he referred to Mumbai's firefighters as "angels from Heaven." As fire crews finally reached Diffenderffer, he said they surrounded him in order to protect him from gunfire.

"There were some pretty scary moments while we were fighting fires at the Taj Hotel. On Thursday night, while we were in a cage trying to rescue guests trapped in the heritage section of the hotel, we saw a gun-toting terrorist," A.V. Sawant, Chief Fire Officer, said.

"Luckily for us, he did not turn in our direction and we continued fighting the fire," Mr. Sawant said. The fire brigade was called in to battle fires multiple times at the Taj and Oberoi hotels.

"They [the terrorists] kept setting the rooms on fire while fighting the commandos. We would wait for an assurance from the commandos before going in," Mr. Sawant said.

However, the sound of firing did not make it easy for them, he said. "It was not easy to work in such conditions. There is a risk while fighting fires but in this case there was the added risk of bullets," a fireman said.

In the wake of the late November 2008 events, it was revealed that Mumbai's firefighters have poor personal protective gear. One fireman best described the brigades' Personal protective equipment (PPE) as similar to London's circa 1950. Members currently wear wool tunics and compressed cork helmets. State-of-the-art firefighting gear has apparently been authorised by the Municipal Corporation of Greater Mumbai but, 13 months later, front-line firefighting teams had yet to see these lifesaving garments.

Contact numbers

Hotline 101 (Unmetered Number)

Trivia

Majority of Fire Station has Ganesha temple. Because lord Ganesha is associated with coolness, and clearing obstacles. It is believed that due to cool nature of Ganesha it will shield Firefighter from heat of fire.

See also

- Bombay Explosion (1944)
- Mumbai Police
- History of Mumbai

Gallery

Mumbai Frie Brigade's Officer oprating Telescoping articulated platform

Fire Brigade's Modern Aerial Lift with a attendant

Close up of Hydraulic Platform

Telescoping hydraulic platform Remote

References

http://www.maharashtra.gov.in/english/gazetteer/greater_bombay/local.html#1

[1] http://www.change.godrej.com/septoct/touch_heroism.htm

[2] http://www.mcgm.gov.in/irj/portalapps/com.mcgm.ahome_keyprojects/docs/2-8%20Fire%20Services.pdf

[3] http://www.change.godrej.com/septoct/touch_heroism.htm

[4] http://books.google.co.uk/books?id=nyokqAQOgMsC&printsec=frontcover#v=onepage&q=&f=false

[5] http://midcfireservice.com/dailyroutine.html

[6] http://www.change.godrej.com/septoct/touch_heroism.htm

[7] http://www.maharashtrafireservices.org/fire_service_week.htm

Karachi

Karachi

کراچی

ڪراچي

— **City District** —

Clockwise from top left: French Beach, Karachi, Merewether Clock Tower, D. J. Science College, Karachi Creek Marina, MCB Tower, Mazar-e-Quaid.

Logo

Nickname(s): The Gateway to Pakistan, The City of Lights

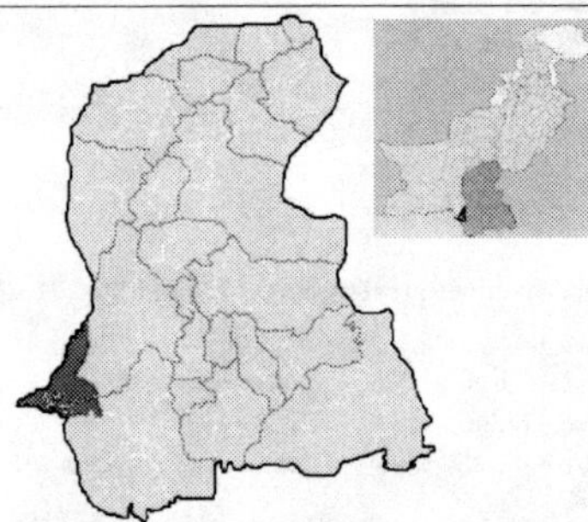

Location of Karachi in Sindh and in Pakistan

Karachi

Location of Karachi in Pakistan

Coordinates: 24°51′36″N 67°0′36″E	
Country	Pakistan
Province	Sindh
Municipal Committee	1853
Municipal Corporation	1933
Metropolitan Corporation	1976
City District Government	14th August 2001
City Council	City Complex, Gulshan Town
Towns	
Government [1]	
- Type	City District
- City Nazim	Syed Mustafa Kamal
- Naib Nazim	Nasreen Jalil
- DCO	Javed Hanif
Area [2]	
- Total	3530 km^2 (1362.9 sq mi)
Elevation	8 m (26 ft)
Population (2009)[3] [4]	
- Total	between 12 and 18 million
Time zone	PST (UTC+5)
Area code(s)	021
Website	http://www.karachicity.gov.pk

Karachi (Urdu: کراچی, Sindhi: راچي) is the largest city, main seaport and the financial capital of Pakistan, and the capital of the province of Sindh. With a city population of 15.5 million, Karachi is one of the world's largest cities,[4] 13th largest urban agglomeration[5] , the 20th largest metropolitan area in the world,[6] and the 2nd largest city within the Organisation of the Islamic Conference. It is Pakistan's premier centre of banking, industry, and trade. Karachi is home to Pakistan's largest corporations, including those that are involved in textiles, shipping, automotive industry, entertainment, the arts, fashion, advertising, publishing, software development and medical research. The city is a major hub of higher education in South Asia and the wider Islamic world.[7] Karachi is ranked as a Beta world city.[8] [9]

Karachi enjoys its prominent position because of its geographical location on a bay, making it the financial capital of the country. It is one of the fastest growing cities in the world. It was the original capital of Pakistan until the construction of Islamabad and is the location of the Port of Karachi and Port Bin Qasim, one of the region's largest and busiest ports. After the partition of India and the independence of Pakistan, the city's population increased dramatically when hundreds of thousands of migrants from India, East Pakistan (later Bangladesh) and other parts of South Asia came to settle in the city.

Karachi city is spread over 3530 km^2 (1360 sq mi) in area, almost five times bigger than Singapore. It is locally known as the "City of Lights" (رھش وج ںینشور) and "The bride of the cities" (سورع دالبلا) for its liveliness, and the "City of the Quaid" (رھش ِدئاق), having been the birth and burial place of Quaid-e-Azam (Muhammad Ali Jinnah), the founder of Pakistan, who made the city his home after Pakistan's independence.

History

Sindh Arts College, now renamed as D. J. Science College

The area of Karachi was known to the ancient Greeks by many names: Krokola, the place where Alexander the Great camped to prepare a fleet for Babylonia after his campaign in the Indus Valley; 'Morontobara' (probably Manora island near Karachi harbour), from whence Alexander's admiral Nearchus set sail; and Barbarikon, a port of the Indo-Greek Bactrian kingdom. It was later known to the Arabs as Debal, the starting point for Muhammad bin Qasim and his army in 712 AD. Karachi was founded as "Kolachi" by Baloch tribes from Balochistan and Makran, who established a small fishing community in the area.[10] Descendants of the original community still live in the area on the small island of Abdullah Goth, which is located near the Karachi Port. The original name "Kolachi" survives in the name of a well-known Karachi locality named "Mai Kolachi". The city was visited by Ottoman Admiral Seydi Ali Reis in 1550s and mentioned in his book *Mirat ul Memalik* (The Mirror of Countries), 1557 AD.[11] The present city started life as a fishing settlement when a Balochi fisherwoman called Mai Kolachi took up residence and started a family. The village that later grew out of this settlement was known as *Kolachi-jo-Goth* (Village of Kolachi in Sindhi). By the late 1720s, the village was trading across the Arabian Sea with Muscat and the Persian Gulf region. A small fort was constructed for its protection, armed with cannons imported from Muscat. The fort had two main gateways: one facing the sea, known as Kharra Darwaaza (Brackish Gate) (Kharadar) and the other facing the Lyari River known as the Meet'ha Darwaaza (Sweet Gate) (Mithadar).[12] The location of these gates correspond to the modern areas of Kharadar (*Khārā Dar*) and Mithadar (*Mīṭhā Dar*).

A view of Saddar Bazaar in 1900

After sending a couple of exploratory missions to the area, the British East India Company conquered the town when HMS *Wellesley* anchored off Manora island on 1 February 1839. Two days later, the little fort surrendered.[13] The town was later annexed to the British Indian Empire when Sindh was conquered by Charles James Napier in Battle of Miani on 17 February 1843. On his departure in 1847, he is said to have remarked, "Would that I could come again to see you in your grandeur!" Karachi was made the capital of Sindh in the 1840s. On Napier's departure, it was added along with the rest of Sindh to the Bombay Presidency, a move that caused considerable resentment among the native Sindhis. The British realised the importance of the city as a military cantonment and as a port for exporting the produce of the Indus River basin, and rapidly developed its harbour for shipping. The foundations of a city municipal government were laid down and infrastructure development was undertaken. New businesses started opening up and the population of the town began rising rapidly. The arrival of the troops of the Kumpany Bahadur in 1839 spawned the foundation of the new section, the military cantonment. The cantonment formed the basis of the 'white' city, where the Indians were not allowed free access. The 'white' town was modeled after English industrial parent-cities, where work and residential spaces were separated, as were residential from recreational places. Karachi was divided into two major poles. The 'black' town in the northwest, now enlarged to accommodate the burgeoning Indian mercantile population. When the Indian Rebellion of 1857 broke out in South Asia, the 21st Native Infantry, then stationed in Karachi, declared allegiance to rebels and joining their numbers on 10 September 1857. Nevertheless, the British were able to quickly reassert control over Karachi and defeat the uprising.

Karachi Airport in 1943 during World War II

In 1864, the first telegraphic message was sent from India to England, when a direct telegraph connection was laid between Karachi and London.[14] In 1878, the city was connected to the rest of British India by rail. Public building projects, such as Frere Hall (1865) and the Empress Market (1890), were undertaken. In 1876, Muhammad Ali Jinnah, the founder of Pakistan, was born in the city, which by now had become a bustling city with mosques, churches, courthouses, kota, paved streets and a magnificent harbour. By 1899, Karachi had become the largest wheat exporting port in the East.[15] The original population of Karachi consisted of Balochis and Sindhis. Karachi was a small port town and part of Talpur dynasty in Sindh. The British East India Company conquered Karachi on February 3, 1839 and started developing it as a major port town. These developments in Karachi resulted in large influx of economic migrants: Parsis, Hindus, Christians, Jews, Marathis, Goans, Armenians, Chinese, British, Lebanese and Gujaratis. The population of the city was about 105,000 inhabitants by the end of the 19th century, with a cosmopolitan mix different nationalities. British colonialists embarked on a number of public works of sanitation and transportation — such as gravel paved streets, proper drains, street sweepers, and a network of trams and horse-drawn trolleys. Colonial administrators set up military camps, a European inhabited quarter, and organised marketplaces, of which the Empress Market is most notable.

By the time the new country of Pakistan was formed in 1947 Karachi had become a bustling metropolis with beautiful classical and colonial European styled buildings, lining the city's thoroughfares. Karachi was chosen as the capital of Pakistan, which at the time included modern day Bangladesh, a region located more than 1000 km (620 mi) away, and not physically connected to Pakistan. In 1947, Karachi was the focus for settlement by Muslim migrants from India, who drastically expanded the city's population and transformed the demographics and economy. In 1958, the capital of Pakistan was moved from Karachi to Rawalpindi and then in 1960, to the newly built Islamabad. This marked the start of a long period of decline in the city, marked by a lack of development.[16] Karachi had both a municipal corporation and a Karachi Divisional Council in the 1960s, which developed plans for schools, colleges, roads, municipal gardens, and parks. The Karachi Divisional Council had separate working committees for education, roads, and residential societies development and planning.[17] During the 1960s, Karachi was seen as an economic role model around the world. Many countries sought to emulate Pakistan's economic planning strategy and one of them, South Korea, copied the city's second "Five-Year Plan" and World Financial Centre in Seoul is designed and modeled after Karachi.[18] [19]

The 1970s saw major labour struggles in Karachi's industrial estates, (see: Karachi labour unrest of 1972). The 1980s and 1990s saw an influx of refugees from the Soviet war in Afghanistan into Karachi, they were followed in smaller numbers by refugees escaping from Iran.[20] Political tensions between the Muhajir and other *native* groups (e.g.

Sindhis, Punjabis, Pashtuns, and others), erupted and the city was wracked with political and ethnic violence. The period from 1992 to 1994 is regarded as the bloodiest period in the history of the city, when the Army commenced its Operation Clean-up against the Muttahida Qaumi Movement. Most of these tensions have now simmered down. Today, Karachi continues to be an important financial and industrial centre and handles most of the overseas trade of Pakistan and the other Central Asian countries. It accounts for a lion's share of the GDP of Pakistan,[21] and a large proportion of the country's white collar workers.[22] The multi-ethnic mix can be imagined from the fact that there are more Pashtuns in Karachi than in any city of the North-West Frontier Province.

Geography

Karachi is located in the south of Pakistan, on the coast of the Arabian Sea. Its geographic coordinates are 24°51′ N 67°02′ E. Most of the land comprised largely of flat or rolling plains, with hills on the western and Manora Island and the Oyster Rocks. The Arabian Sea beach lines the southern coastline of Karachi. Mangroves and creeks of the Indus delta can be found toward the southeast side of the city. Toward the west and the north is Cape Monze, locally known as Raas Muari, an area marked by projecting sea cliffs and rocky sandstone promontories. Some excellent beaches can be found in this area. Khasa Hills lie in the north west and form the border between North Nazimabad Town and Orangi Town. The Manghopir mountain range lies in north west of Karachi and between Hub River and Manghopir.

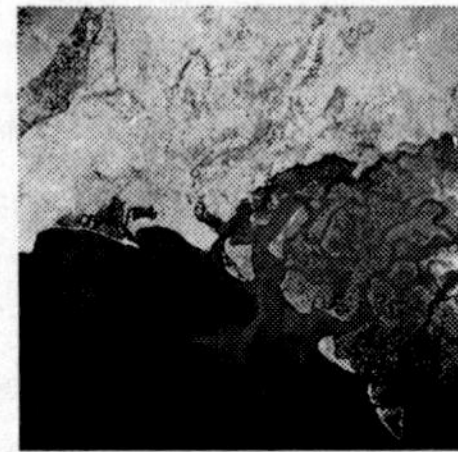

Satellite view of Karachi

Karachi's Oyster Rocks near Clifton Beach

Climate

Evening scene on Clifton beach.

Located on the coast, Karachi tends to have a relatively mild, arid climate with low average precipitation levels (approximately 250 mm per annum), the bulk of which occurs during the July-August monsoon season. Winters are mild and the summers are hot; the proximity to the sea maintains humidity levels at a near-constant high and cool sea breezes relieve the heat of the summer months. Because of high temperatures during the summer (ranging from 30 to 44 degrees Celsius from April to August), the winter months (November to February) are generally considered the best times to visit Karachi. December and January have pleasant and cloudy weather.

In 2003, 2006, 2007 and 2009 Karachi was affected by heavy to extremely heavy rainfall. On June 23, 2007, Cyclone Yemyin lashed the city with heavy downpours and strong windstorms. On July 18, 2009, there was severe flooding, in which a record-breaking rainfall of 235 mm occurred in just 14 hours, killing 20 and injuring 150 people. The city's highest monthly rainfall—711.5 mm (28.0) inches—occurred in July 1869. Karachi's highest recorded temperature is 48 °C (118 °F) and its lowest is 0.0 °C (32 °F).[23]

Climate data for Karachi, Pakistan

Month	Jan	Feb	Mar	Apr	May	Jun	Jul	Aug	Sep	Oct	Nov	Dec	Year
Record high °C (°F)	32 (90)	34 (93)	42 (108)	44 (111)	48 (118)	46 (115)	43 (109)	37 (99)	41 (106)	42 (108)	38 (100)	33 (91)	48 (118)
Average high °C (°F)	25 (77)	26 (79)	29 (84)	32 (90)	34 (93)	34 (93)	33 (91)	31 (88)	31 (88)	33 (91)	31 (88)	27 (81)	30.5 (87)
Average low °C (°F)	13 (55)	14 (57)	19 (66)	23 (73)	26 (79)	28 (82)	27 (81)	26 (79)	25 (77)	22 (72)	18 (64)	14 (57)	21.3 (70)
Record low °C (°F)	0 (32)	3 (37)	7 (45)	12 (54)	18 (64)	22 (72)	22 (72)	20 (68)	18 (64)	10 (50)	6 (43)	1 (34)	0 (32)
Precipitation mm (inches)	13 (0.51)	10 (0.39)	8 (0.31)	3 (0.12)	3 (0.12)	18 (0.71)	85 (3.35)	61 (2.4)	13 (0.51)	0 (0)	3 (0.12)	5 (0.2)	222 (8.74)
Source: [24] & [25] *5.08.2009*													

Demographics

Population growth			
Census	**Pop.**		**%±**
1881	73560		—
1891	105199		43.0%
1901	136297		29.6%
1911	186771		37.0%
1921	244162		30.7%
1931	300799		23.2%
1941	435887		44.9%
1951	1068459		145.1%
1961	1912598		79.0%
1972	3426310		79.1%
1981	5208132		52.0%
1998	9339023		79.3%
Est. 2009	12-18000000		—

Source:[3]
† Huge population rise in 1951 because of large
scale migration after Partition of India in 1947.

The population and demographic distribution in Karachi has undergone numerous changes over the past 150 years. The original population of Karachi consisted of Balochis and Sindhis. Karachi was a small port town and part of Talpur dynasty in Sindh. The British East India Company conquered Karachi on February 3, 1839 and started developing it as a major port town. These developments in Karachi resulted in large influx of economic migrants: Parsis, Hindus, Christians, Jews, Marathis, Goans, Armenians, Chinese, British, Lebanese and Gujaratis. The population of the city was about 105,000 inhabitants by the end of the 19th century, with a cosmopolitan mix of different nationalities. Non-governmental and international estimates of Karachi's population run anywhere from 12 million to 18 million[3] [26] [27] – a huge increase over its population in 1947 (400,000). It is estimated that over 90% of its population are migrants from different backgrounds. The city's population is growing at about 5% per year (mainly as a result of rural-urban internal migration), including an estimated 45,000 migrant workers coming to the city every month from different parts of Pakistan.[28]

After independence of Pakistan, a large number of Hindus and Sikhs left the city for Gujarat, Rajasthan, and Punjab, although a large Hindu community still exists in Karachi, as is evident by numerous functioning Hindu Temples. Most Hindu properties were inhabited by Muslim migrants known as Muhajirs who escaped from the pogroms and genocide in India and settled in Pakistan after independence in 1947. The Muhajirs migrated from different parts of India; the majority of them spoke Urdu. Muhajirs now make up a majority of Karachi's residents while Pashtuns are the city's second-largest ethnic group.[29] [30] Karachi has a cosmopolitan mix of many ethno-linguistic groups from all over Pakistan and refugees from neighboring countries.[31] The Sheedis, the local name for Afro-Pakistanis, trace their roots to African slaves.[32]

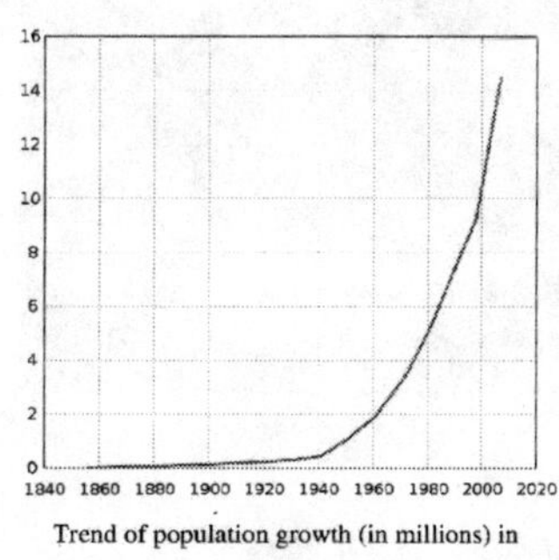

Trend of population growth (in millions) in Karachi

After the Indo-Pakistani War of 1971, thousands of Biharis and Bengalis from Bangladesh arrived in the city, and today, Karachi is home to 1 to 2 million ethnic Bengalis from Bangladesh,[33] [34] many of whom immigrated in the 1980s and 1990s and work as fishermen. These immigrants were followed by Rohingya refugees[35] from Burma, other Burmese Muslims and Asians from Uganda. According to the UNHCR and the local law enforcement, approximately 50,000 registered Afghan refugees live in Karachi.[36] Many other refugees from Iran and Tajikistan have settled in the city. Tens of thousands of Central Asians from the former Soviet Union and Nepalis have also settled in the city as economic migrants. There also exists a economic elite of Sinhalese from Sri Lanka. With up to 7 million Pashtuns by some estimates,[29] Karachi hosts the largest Pashtun population in the world, far outnumbering the number of Pashtuns in Peshawar, Quetta, Kandahar, or Kabul — despite Karachi's distance of over a thousand kilometers from the Pashtun heartlands in Afghanistan, northern Baluchistan, and Khyber-Pakhtunkhwa. Many of these Pashtuns have been resident in Karachi for decades, and as a result, many no longer speak Pashto fluently, and instead primarily speak Urdu or English — especially those from wealthier communities. In addition, a large portion of the Muhajirs community in Karachi claim to be ethnic Pashtuns.

According to the census of Pakistan 1998, the religious breakdown of the city is as follows:[37] Muslim (96.45%), Christian (2.42%), Hindu (0.86%), Ahmadi (0.17%) and other (0.10%). (Other religious groups include Parsis, Sikhs, Bahai, Jews and Buddhists.)

The most commonly spoken language in Karachi is Urdu, the national language. Other national languages spoken in Karachi are Sindhi, Punjabi, Pashto and Balochi are widely spoken in the city. As per the old census of Pakistan in 1998, the distribution of national languages within the city was:[37] Urdu (48.52%), Punjabi (13.94%), Pashto (11.42%), Sindhi (7.22%), Balochi (4.34%), Saraiki (2.11%) and other (12.44%). However, these numbers only include major national languages of Pakistan, and neglect migrant languages, some of which such as Bengali and Dari are widely spoken by a high number of residents. Other languages with a significant number of speakers include Arabic, Brahui, Burushaski, Gujarati, Hindko, Khowar, Makrani, Marathi, Jatki, Memoni, Persian, and Sinhalese. Numerous other languages are spoken by small communities within the city.

Economy

Karachi is the financial and commercial capital of Pakistan. In line with its status as a major port and the country's largest metropolis, it accounts for a lion's share of Pakistan's revenue. According to the Federal Board of Revenue's 2006-2007 year book, tax and customs units in Karachi were responsible for 46.75% of direct taxes, 33.65% of federal excise tax, and 23.38% of domestic sales tax.[38] Karachi accounts for 75.14% of customs duty and 79% of sales tax on imports.[38] Therefore, Karachi collects a significant 53.38% of the total collections of the Federal Board of Revenue (since renamed as the Central Board of Revenue), out of which 53.33% are

Korangi Road

customs duty and sales tax on imports.[38] (Note: Revenue collected from Karachi includes revenue from some other areas since the Large Tax Unit (LTU) Karachi and Regional Tax Offices (RTOs) Karachi, Hyderabad, Sukkur & Quetta cover the entire province of Sindh and Balochistan).[38] Karachi's indigenous contribution to national revenue is 25%.[21]

I. I. Chundrigar Road

Karachi's contribution to Pakistan's manufacturing sector amounts to approximately 30 percent.[39] A substantial chunk of Sindh's gross domestic product (GDP) is attributed to Karachi[40] [41] (the GDP of Sindh as a percentage of Pakistan's total GDP has traditionally hovered around 28%-30%).[40] [41] [42] [43] Karachi's GDP is around 20% of the total GDP of Pakistan.[21] [44] A PricewaterhouseCoopers study released in 2009, which surveyed the 2008 GDP of the top cities in the world, calculated Karachi's GDP (PPP) to be $78 billion[45] (projected to be $193 billion in 2025 at a growth rate of 5.5%).[45] It confirmed Karachi's status as Pakistan's largest economy, well ahead of the next two biggest cities Lahore and Faisalabad, which had a reported GDP (PPP) in 2008 of $40 billion and $14 billion, respectively.[45] Karachi's high GDP is based on its mega-industrial base, with a high dependency on the financial sector. Textiles, cement, steel, heavy machinery, chemicals, food, banking and insurance are the major segments contributing to Karachi's GDP. In February 2007, the World Bank identified Karachi as the most business-friendly city in Pakistan.[46]

Shah-re-Faisal Road, a part of Karachi's financial district

The MCB Tower and downtown Karachi

Karachi is the nerve center of Pakistan's economy. The economic stagnation caused by political anarchy, ethnic strife and resultant military operation during late 1980s and 90s led to efflux of industry from Karachi. Most of Pakistan's public and private banks are headquartered on Karachi's I. I. Chundrigar Road; according to a 2001 report, nearly 60% of the cashflow of the Pakistani economy takes place on I. I. Chundrigar Road. Most major foreign multinational corporations operating in Pakistan have their headquarters in Karachi. The Karachi Stock Exchange is the largest stock exchange in Pakistan, and is considered by many economists to be one of the prime reasons for Pakistan's 8% GDP growth across 2005.[47] A recent report by Credit Suisse on Pakistan's stock market is a testimonial to its strong fundamentals, estimating Pakistan's relative return on equities at 26.7 percent, compared to Asia's 11 percent.[48]

Recently, Karachi has seen an expansion of information and communications technology and electronic media and has become the software outsourcing hub of Pakistan. Call centres for foreign companies have been targeted as a significant area of growth, with the government making efforts to reduce taxes by as much as 10% in order to gain foreign investments in the IT sector.[49] [50] Many of Pakistan's independent television and radio stations are based in Karachi, including world-popular Business Plus, AAJ News, Geo TV, KTN,[51] Sindh TV,[52] CNBC Pakistan, TV ONE, ARY Digital, Indus Television Network, Samaa TV and Dawn News, as well as several local stations.

Karachi has several large industrial zones such as Karachi Export Processing Zone, SITE, Korangi, Northern Bypass Industrial Zone, Bin Qasim and North Karachi, located on the fringes of the main city.[53] Its primary areas of industry are textiles, pharmaceuticals, steel, and automobiles. In addition, Karachi has a vibrant cottage industry and there is a rapidly flourishing Free Zone with an annual growth rate of nearly 6.5%. The Karachi Expo Centre hosts many regional and international exhibitions.[54] There are many development projects proposed, approved and under construction in Karachi. Among projects of note, Emaar Properties is proposing to invest $43bn (£22.8bn) in Karachi to develop Bundal Island, which is a 12,000 acre (49 km²) island just off the coast of Karachi.[55] The Karachi Port Trust is planning a Rs. 20 billion, 1947 feet (593 m) high Port Tower Complex on the Clifton shoreline.[56] [57] It will comprise a hotel, a shopping center, an exhibition center and a revolving restaurant with a viewing gallery offering a panoramic view of the coastline and the city.

Arts and culture

Karachi is home to some of Pakistan's important cultural institutions. The National Academy of Performing Arts,[58] located in the newly renovated Hindu Gymkhana, offers a two-year diploma course in performing arts that includes classical music and contemporary theatre. The All Pakistan Music Conference, linked to the 45-year-old similar institution in Lahore, has been holding its Annual Music Festival since its inception in 2004. The Festival is now a well-established feature of the city life of Karachi that is attended by more than 3000 citizens of Karachi as well as people from other cities.[59] The National Arts Council (*Koocha-e-Saqafat*) has musical performances and mushaira (poetry recitations). The Kara Film Festival annually showcases independent Pakistani and international films and documentaries.

Mohatta Palace

National Museum of Pakistan

Karachi has many museums that present exhibitions on a regular basis, including the Mohatta Palace and the National Museum of Pakistan. Karachi Expo Centre hosts many regional and international exhibitions.

The everyday lifestyle of Karachi differs substantially from that of other Pakistani cities and towns. The culture of Karachi is characterized by the blending of South Asian, Middle Eastern, Central Asian and Western influences, as well as its status as a major international business centre. After the partition of Indian subcontinent, Karachi received a large number of refugees from all over India, whose influence is now evident in the city's different sub-cultures. Karachi hosts the largest middle class stratum of the country.

Architecture

Frere Hall, Karachi

Karachi has a rich collection of buildings and structures of various architectural styles. Many modern high-rise buildings are currently under construction. The downtown districts of Saddar and Clifton contain a variety of early 20th-century architecture, ranging in style from the neo-classical KPT building to the Sindh High Court Building. During the period of British rule, classical architecture was preferred for monuments of the British Raj. Karachi acquired its first neo-Gothic or Indo-Gothic buildings when Frere Hall, Empress Market and St. Patrick's Cathedral were completed. The Mock Tudor architectural style was introduced in the Karachi Gymkhana and the Boat Club. Neo-Renaissance architecture was popular in the 19th century and was the language for St. Joseph's Convent (1870) and the Sind Club (1883).[60] The classical style made a comeback in the late nineteenth century, as seen in Lady Dufferin Hospital (1898)[61] and the Cantt. Railway Station. While Italianate buildings remained popular, an eclectic blend termed Indo-Saracenic or Anglo-Mughal began to emerge in some locations.

The local mercantile community began acquiring impressive mercantile structures. Zaibunnisa Street in the Saddar area (known as Elphinstone Street in British days) is an example where the mercantile groups adopted the Italianate and Indo-Saracenic style to demonstrate their familiarity with Western culture and their own. The Hindu Gymkhana (1925) and Mohatta Palace are the example of Mughal revival buildings.[62] The Sindh Wildlife Conservation Building, located in Saddar, served as a Freemasonic Lodge until it was taken over by the government. There are talks of it being taken away from this custody and being renovated and the Lodge being preserved with its original woodwork and ornate wooden staircase.[63]

In recent years, a large number of architecturally distinctive, even eccentric, buildings have sprung up throughout Karachi. Notable examples of contemporary architecture include the Pakistan State Oil Headquarters building and the Karachi Financial Towers. The city has numerous examples of modern Islamic architecture, including the Aga Khan University hospital, Masjid e Tooba, Faran Mosque, Bait-ul Mukarram Mosque, Quaid's Mausoleum, and the Textile Institute of Pakistan. One of the unique cultural elements of Karachi is that the residences, which are two- or three-story townhouses, are built with the front yard protected by a high brick wall. Ibrahim Ismail Chundrigar Road features a range of extremely tall buildings. The most prominent examples include the Habib Bank Plaza, PRC Towers and the MCB Tower which is the tallest skyscraper in Pakistan.[64] A spectacular skyscraper, Port Tower Complex, has been proposed in the Clifton District of the metropolis. At 593 metres, the building will comprise a hotel, a shopping centre, an exhibition centre and a revolving restaurant with a viewing gallery offering a panoramic view of the coastline and the city.[65]

Many more high-rise buildings are under construction, such as Centre Point near Korangi Industrial Area, IT Tower, Sofitel Tower Karachi and Emerald Tower. The Government of Sindh recently approved the construction of two high-density zones, which will host the new city skyline.

Fashion, shopping and entertainment

Karachi has always been proactive in organizing large events but because of the political and economic crisis in the country, activities have recently been slowed down. Karachi continues to host many different cultural and fashion shows. In 2009 a four-day-long fashion show was organized in Karachi's luxury Marriott hotel.[66] Karachi has many glitzy shopping malls in the Clifton area, Tariq Road, Gulshan-e-Iqbal and Hyderi shopping area, such as Park Towers, The Forum, Dolmen Mall and Millenium Mall. Zamzama Boulevard is known for its designer stores and many cafes. There are many bazaars in Karachi selling different merchandise. The famous bazaars include: Bohri

Bazaar, Soldier Bazaar, Urdu Bazaar, etc. Foreign clothes brands and famous Pakistani fashion labels (such as Amir Adnan, Aijazz, Rizwan Beyg, Deepak Perwani, Shayanne Malik, Maria B, Khaadi, Sputnik Footwear, Metro Shoes, English Boot House, Cotton & Cotton, Men's Store and Junaid Jamshed) are present in various major shopping districts of the city.

Sports

National Stadium, Karachi

Cricket is the most popular sport of the city and is usually played in many small grounds around the city. Gully cricket is played in the narrow by-lanes of the city. Nighttime cricket, popularly called 'night match', can be seen at weekends when people play brightly lit, late night matches on less traversed city streets and the game continues till dawn. Karachi has one world-class cricket stadium, named National Stadium. The National Stadium (NSK), the largest cricket stadium in Pakistan, became Karachi's fifth and Pakistan's 11th first-class ground. The inaugural first-class match was played at NSK between Pakistan and India on April 21–24, 1955. In 34 Tests between that first match and December 2000, Pakistan won 17 and were never beaten. Their only Test defeat on the ground came in the gloom against England in 2000-01. Since then, major terrorist activity, mainly bombings, have meant that non-Asian sides have refused to play in the city, and in five years only Bangladesh and Sri Lanka have visited.

The first One Day International at the National Stadium was against West Indies on November 21, 1980, and it went down to the last ball as Gordon Greenidge drove Imran Khan imperiously to the cover boundary with three needed. It has been a far less successful limited-overs venue, with defeats outnumbering victories. In fact, in a little under five years from the start of 1996, Pakistan failed to win on the ground. It staged a quarter-final match in the 1996-97 World Cup.

Other popular sports in the city are hockey, boxing, association football, golf, table tennis, snooker, squash, and horse racing. Sports like badminton, volleyball and basketball are popular in schools and colleges. Football is especially popular in Lyari Town, which has a large Afro-Balochi community and has always been a football-mad locality in Karachi. The Peoples Football Stadium is perhaps the largest football stadium in Pakistan with respect to capacity, easily accommodating around 40,000 people. In 2005, the city hosted the SAFF Championship at this ground, as well as the Geo Super Football League 2007, which attracted capacity crowds during the games.

The city has facilities for hockey (the Hockey Club of Pakistan, UBL Hockey Ground), boxing (KPT Sports Complex), squash (Jahangir Khan Squash Complex) and polo. Marinas and boating clubs add to the diverse sporting activities in Karachi.

Government

Civic Centre. Head Office of the City District Government, Karachi

The City of Karachi Municipal Act was promulgated in 1933. Initially, the Municipal Corporation comprised the mayor, the deputy mayor and 57 councillors. The Karachi Municipal Corporation was changed to a Metropolitan Corporation in 1976. The administrative area of Karachi was a second-level subdivision known as Karachi Division, which was subdivided into five districts: Karachi Central, Karachi East, Karachi South, Karachi West and Malir. In 2000, the national government implemented a new devolution plan which abolished the second-tier divisions and merged the five districts of Karachi into a new City District, structured as a three-tiered federation, with the two lower tiers composed of 18 towns and 178 union councils (UC).[67]

The towns are governed by elected municipal administrations responsible for infrastructure and spatial planning, development facilitation, and municipal services (water, sanitation, solid waste, repairing roads, parks, street lights, and traffic engineering), with some functions being retained by the City-District Government (CDG).[67] The third-tier 178 union councils are each composed of thirteen directly elected members including a Nazim (mayor) and a Naib Nazim (deputy mayor). The UC Nazim heads the union administration and is responsible for facilitating the CDG to plan and execute municipal services, as well as for informing higher authorities about public concerns and complaints. Naimatullah Khan was the first Nazim of Karachi and Shafiq-Ur-Rehman Paracha was the first DCO of Karachi, Paracha even served as the last Commissioner of Karachi. Naimatullah Khan focused on building new parks, providing entertainment outlets to the youth (to celebrate events like Valentine's Day) and families (to celebrate events like Eid).

In the elections of 2005, Mustafa Kamal was elected City Nazim of Karachi to succeed Naimatullah Khan, and Nasreen Jalil was elected as the City Naib Nazim. Mustafa Kamal was previously the provincial minister for information technology in Sindh. Mustafa Kamal is advancing the development trail and has been actively involved in maintaining care of the city's municipal systems.[68]

There are six military cantonments administered by the Pakistan Army which do not form part of the City District Government of Karachi.

1. Lyari Town		1. Liaquatabad Town
2. Saddar Town		2. North Nazimabad Town
3. Jamshed Town		3. Gulberg Town
4. Gadap Town		4. New Karachi Town
5. SITE Town		5. Orangi Town
6. Kemari Town		6. Baldia Town
7. Shah Faisal Town		A. Karachi Cantonment
8. Korangi Town		B. Clifton Cantonment
9. Landhi Town		C. Korangi Creek Cantonment
10. Bin Qasim Town		D. Faisal Cantonment
11. Malir Town		E. Malir Cantonment
12. Gulshan Town		F. Manora Cantonment

Education

In 2008-09, the city's literacy rate was estimated at 65.26%,[69] the highest in Pakistan,[70] with a gross enrolment ratio of 111%, the highest in Sindh.[71]

Education in Karachi is divided into five levels: primary (grades one through five); middle (grades six through eight); high (grades nine and ten, leading to the Secondary School Certificate); intermediate (grades eleven and twelve, leading to a Higher Secondary School Certificate); and university programs leading to graduate and advanced degrees. Karachi has both public and private educational institutions. Most educational institutions are gender-based, from primary to university level.

National University of Sciences and Technology, Karachi Campus

Karachi Grammar School is the oldest school in Pakistan and has educated many Pakistani businessmen and politicians. The Narayan Jagannath High School in Karachi, which opened in 1855, was the first government school established in Sindh. Other well-known schools include the British Overseas School, L'ecole for Advanced Studies, Generation's school, the CAS School, Bay View High School, Bay View Academy, Karachi American School, Aga Khan Higher Secondary School, The Paradise School and College, Little Folks Secondary School, Habib Public School, Mama Parsi Girls Secondary School, B. V. S. Parsi High School, Civilizations Public School, The Oasys School, Avicenna School, The Lyceum School, Ladybird Grammar School, The City School, Beaconhouse School System, The Educators schools, Shahwilayat Public School, St Patrick's High School, St Paul's English High School, St Joseph's Convent School, St Jude's High School, St Michael's Convent School and Foundation Public School.

The University of Karachi, known as KU, is Pakistan's largest university, with a student population of 24,000 and one of the largest faculties in the world. It is located next to the NED University of Engineering and Technology, the country's oldest engineering institute. In the private sector, Sir Syed University of Engineering and Technology (SSUET) provides reputable training in biomedical engineering, civil engineering, electronics engineering, telecom engineering and computer engineering. Dawood College of Engineering and Technology, which opened in 1962, offers degree programmes in electronic engineering, chemical engineering, industrial engineering, materials engineering and architecture. Karachi Institute of Economics & Technology (KIET) has two campuses in Karachi and has been growing rapidly since its inception in 1997. The Plastics Technology Center (PTC), located in Karachi's Korangi Industrial Area, is at present Pakistan's only educational institution providing training in the field of polymer engineering and plastics testing services.[72] The Institute of Business Administration (IBA), founded in 1955, is the oldest business school outside of North America. The Shaheed Zulfiqar Ali Bhutto Institute of Science and Technology (SZABIST), founded in 1995 by Benazir Bhutto, is located in Karachi, with its other campuses in Islamabad, Larkana and Dubai. Pakistan Navy Engineering College (PNEC) is a part of the National University of Sciences and Technology (NUST), offering a wide range of engineering programs, including electrical engineering and mechanical engineering. Hamdard University is the largest private university in Pakistan. Karachi is home of the head offices of the Institute of Chartered Accountants of Pakistan (ICAP) (established in 1961) and the Institute of Cost and Management Accountants of Pakistan (ICMAP). Among the many other institutions providing business education are the College of Business Management (CBM), SZABIST, Iqra University and the Institute of Business and Technology (Biztek). Leading medical schools of Pakistan like The Aga Khan University and Dow University of Health Sciences have their campuses in Karachi. The National University of Computer and Emerging Sciences (NUCES-FAST), one of Pakistan's top universities in computer education, operates two campuses in Karachi. PLANWEL [73] is another innovative institution it is a CISCO Network Academy as well as iCBT center for ETS Prometric and Pearsons VUE. Bahria University also has a purpose-built campus in Karachi. The College of Accounting and Management Sciences (CAMS) also has three branches in the city.

For religious education, the Jamia Uloom ul Islamia (one of the largest Islamic education centres of Asia), Jamia Binoria[74] and Darul 'Uloom Karachi are among the Islamic schools in Karachi.

Media

Many of Pakistan's independent television and radio channels are based in Karachi, including Dawn News, Business Plus, Geo TV, CNBC Pakistan, Hum TV, TV ONE, AAJ TV, ARY Digital, Express News, Indus Television Network, Kawish Television Network (KTN), Good News TV, and Sindh TV as well as several local stations. Local channels include Metro One.

Pakistan's premier news television networks are based in Karachi, including News One, GEO News, ARY One World and AAJ News. AAG TV and MTV Pakistan are the main music television channels and Business Plus and CNBC Pakistan are the main business television channels based in the city. The bulk of Pakistan's periodical publishing industry is centred in Karachi, including magazines such as *Spider*, *The Herald*, *Humsay*, *The Cricketer*, *Moorad Shipping News*, and *The Internet*.

Major advertising companies including Interflow Communications and Orient McCann Erickson have their head offices in Karachi.

Crime

Al-Qaeda has a history of using safe houses in Karachi. Ramzi Binalshibh, an al-Qaeda operative described as a "key facilitator for the terrorist attacks on September 11, 2001,"[75] was captured after a gunfight in the city in 2002. Mullah Abdul Ghani Baradar, described as the number two official of the Afghan Taliban, was captured in a "joint" CIA-Pakistani intelligence operation in Karachi in February 2010. Mullah Mohammad Younis, a former Taliban shadow governor in Afghanistan,[76] and Agha Jan Mohtasim, another "Afghan Taliban leader,"[77] were both arrested in the city in early 2010. The Washington Times reported in November 2009 that Mullah Mohammed Omar, the head of the Afghan Taliban, had recently moved to Karachi.[78] Taliban fighters are increasingly using the city to raise money and for vacation.[79] [80] Recent Pakistani media reports even claim Osama bin Laden is hiding in Karachi.[81]

Infrastructure

Transportation

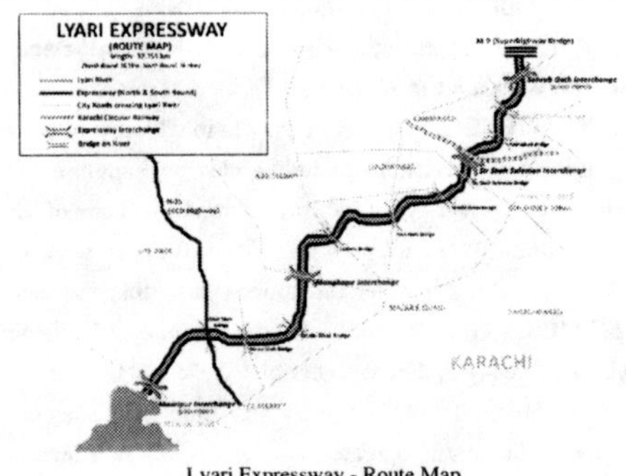

Lyari Expressway - Route Map

The Jinnah International Airport is located in Karachi. It is the largest and busiest airport of Pakistan. It handles 10 million passengers a year. The airport receives the largest number of foreign airlines, a total of 35 airlines and cargo operators fly to Jinnah International predominantly from the Middle East and Southeast Asia. All of Pakistan's airlines use Karachi as their primary transport hub including PIA - Pakistan International Airlines, Airblue, and Shaheen Air International.

The city's old airport terminals are now used for Hajj flights, offices, cargo facilities, and ceremonial visits from heads of state. U.S. Coalition forces used the old terminals for their logistic supply operations as well. The city has two other airstrips, used primarily by the armed forces.

The largest shipping ports in Pakistan are the Port of Karachi and the nearby Port Qasim. These seaports have modern facilities and not only handle trade for Pakistan, but serve as ports for Afghanistan and the landlocked Central Asian countries. Plans have been announced for new passenger facilities at the Port of Karachi.[82]

Jinnah International Airport

Karachi is linked by rail to the rest of the country by Pakistan Railways. The Karachi City Station and Karachi Cantonment Railway Station are the city's two major railway stations. The railway system handles a large amount of freight to and from the Karachi port and provides passenger services to people traveling up country.

A project to transform the existing, but non-operational, Karachi Circular Railway into a modern mass transit system has recently been approved by the government. The $1.6 billion project will be financed by the Japan Bank for International Cooperation and will be completed by 2013. The city government has introduced an initiative to alleviate the transport pains by introducing new CNG buses.

CNG Buses in Karachi

Health and medicine

Karachi is a centre of research in biomedicine. The city is home to at least 30 public hospitals and more than 80 private hospitals, including the Karachi Institute of Heart Diseases, Spencer Eye Hospital, Civil Hospital, PNS Rahat, Abbasi Shaheed Hospital, Aga Khan University Hospital, Holy Family Hospital and Liaquat National Hospital, as well as Jinnah Postgraduate Medical Centre, Ziauddin Hospital, South City Hospital, Kidney Centre, Institute of Child Health, Karachi Institute of Radiology and Nuclear Medicine (KIRAN), Haji Rang Elahi Eye Hospital, Tabba Cardiac Medical Center, Patel Hospital, Layton Benevolent Trust Eye Hospital, Lady Dufferin Hospital, and National Medical Centre.

Challenges of rapid expansion

As one of the most rapidly growing cities in the world, Karachi faces challenges that are central to many developing metropolises, including traffic congestion, pollution, poverty and street crime. These problems continue to earn Karachi low rankings in livability comparisons: *The Economist* ranked Karachi fourth least livable city amongst the 132 cities surveyed[83] and *BusinessWeek* ranked it 175 out of 215 in livability in 2007, down from 170 in 2006.[84]

Traffic problems and pollution are major challenges for Karachi. The level of air pollution in Karachi is estimated to be 20 times higher than World Health Organization standards. A number of new parks (e.g., Bagh Ibne Qasim, Beach View Park and Jheel Park) have been developed and new trees are being planted in the city to improve the environment and reduce the pollution. The construction of new bridges/flyovers, underpasses and signal-free corridors (e.g., Corridor 1: S.I.T.E. to Shahrae Faisal, Corridor 2: North Karachi to Shahrae Faisal, Corridor 3: Safora Goth to Saddar) has improved the traffic flow in Karachi. The eventual completion of Corridor 4 (from the airport to Metropole Hotel) is expected to substantially reduce the travel time to reach the city centre and airport.

Sister cities

Flag	Country	City	District / Region / State	Date
	United States	Houston	Texas	3 March 2009
	Mauritius	Port Louis	Port Louis District	1 May 2007
	China	Shanghai	East China	14 September 2009
	China	Hong Kong	Hong Kong Special Administrative Region	
	Saudi Arabia	Jeddah	Makkah Province	
	Uzbekistan	Tashkent	Tashkent Province	
	Turkey	Istanbul	İstanbul Province	
	Lebanon	Beirut	Beirut Governorate	
	Turkey	İzmir	İzmir Province	since 1985
	United States	New York City	New York	8 May 2008
	Bahrain	Manama	Capital Governorate	28 November 2007
	Kosovo	Pristina	District of Pristina	24 July 2008
	Malaysia	Kuala Lumpur	Federal Territory	1 June 2008
	United States	Washington, D.C.	District of Columbia	since 2008

A twin city partnership with Chicago, Illinois, United States, was contemplated and initiated in 2000, but was never implemented.[85]

There is now an increasing amount of pressure for Karachi and Mumbai to start a twin city partnership because of the many similarities the cities share.[86] [87]

Gallery

Millennium Mall

Empress Market

Korangi Road

Chaukhandi tombs

Karachi Beach

Elphinstone Street c. 1930

I. I. Chundrigar Road

Mohatta Palace

Saint Patrick's Cathedral

Kemari Boat Basin

Fishing boats at the Port of Karachi

TechnoCity Corporate Tower

See also

- Ba'ab-ul-Islam
- List of Sindhi people
- List of cemeteries in Karachi
- List of libraries in Karachi
- List of parks in Karachi
- List of people from Karachi District
- List of places in Karachi
- List of streets of Karachi

Further reading

- The issue of squatters on land needed for new projects [88]
- Karachi: Of encroachments and mega projects [89]

External links

- Karachi City District Government [90]
- History of Karachi [91] A detailed website on historical and modern time on Karachi with photographs
- Karachi FM Radio [92]
- KARACHI - The City of Lights [93]
- Karachi travel guide from Wikitravel
- Karachi [94] at the Open Directory Project
- Mauj Collective for Open Technology Art & Culture_Karachi [95]

Related navpages:

- Navpage: Pakistan topics
- World's most populated metropolitan areas

References

[1] "Government" (http://125.209.91.254/cdgk/Home/Government/tabid/99/Default.aspx). City District Government of Karachi. . Retrieved 2007-11-28.
[2] "About Karachi" (http://125.209.91.254/cdgk/Home/AboutKarachi/tabid/221/Default.aspx). City District Government of Karachi. . Retrieved 2007-11-28.
[3] "The Urban Frontier — Karachi" (http://www.npr.org/templates/story/story.php?storyId=91009748). NPR. June 2, 2008. . Retrieved 2010-01-17. ("...population estimates run anywhere from 12 million to 18 million")
[4] "The largest cities in the world and their mayors" (http://www.citymayors.com/statistics/largest-cities-mayors-1.html). City Mayors. . Retrieved 5 February 2010.
[5] "The world's largest cities and urban areas in 2006" (http://www.citymayors.com/statistics/urban_2006_1.html). City Mayors. . Retrieved 5 February 2010.
[6] R.L. Forstall, R.P. Greene, and J.B. Pick, "Which are the largest? Why published populations for major world urban areas vary so greatly" (http://www.uic.edu/cuppa/cityfutures/papers/webpapers/cityfuturespapers/session3_4/3_4whicharethe.pdf), City Futures Conference, (University of Illinois at Chicago,

July 2004) – Table 5 (p.34)

[7] "Pakistan City Karachi Online Information" (http://pakistancity.org/karachi_online.html). Pakistancity.org. . Retrieved 2010-05-06.
[8] "GaWC - The World According to GaWC 2008" (http://www.lboro.ac.uk/gawc/world2008t.html). Lboro.ac.uk. 2009-06-03. . Retrieved 2009-09-14.
[9] "GAWC World Cities Ranking List" (http://www.diserio.com/gawc-world-cities.html). Diserio.com. . Retrieved 2009-09-14.
[10] R Asif (2002) Lyari Expressway: woes of displaced families (http://www.dawn.com/2002/08/08/fea.htm#2). Dawn (newspaper). 8 August. Retrieved on 10 January 2008
[11] "Mirat ul Memalik" (http://www.fordham.edu/halsall/source/16csidi1.html). Fordham.edu. . Retrieved 2010-05-06.
[12] "History of Karachi" (http://www.shaikhsiddiqui.com/karachi.html). Shaikhsiddiqui.com. 2005-11-10. . Retrieved 2010-05-06.
[13] Neill, , John Martin Bladen (1846). *Recollections of four years' service in the East with H.M. fortieth regiment* (http://www.archive.org/details/recollectionsoff00neilrich). . Retrieved 27 Nov. 2009.
[14] Christina P Harris (1969) The Persian Gulf Submarine Telegraph of 1864. The Geographical Journal (http://www.jstor.org/view/00167398/ap020714/02a00000/0). vol. 135(2). June. pp. 169–190
[15] [Herbert Feldman [1970]: Karachi through a hundred years: the centenary history of the Karachi Chamber of Commerce and Industry 1860–1960. 2. ed. Karachi: Oxford UP (1960).]
[16] "History of Karachi" (http://www.historickarachi.com/1940's.htm). Historickarachi.com. . Retrieved 2010-05-06.
[17] Government archives, Sindh for Municipality and divisional administration
[18] Planning Commission, The Second Five Year Plan: 1960-65, Karachi: Govt. Printing Press, 1960, p. 393
[19] Planning Commission, Pakistan Economic Survey, 1964-65, Rawalpindi: Govt. Printing Press, 1965, p. 212.
[20] "Afghan refugees population in Pakistan - Cambridge Journal" (http://journals.cambridge.org/action/displayAbstract;jsessionid=C8D0B7394F7D074D6832875766C3D91E.tomcat1?fromPage=online&aid=1636848). Journals.cambridge.org. . Retrieved 2010-05-06.
[21] Asian Development Bank. "Karachi Mega-Cities Preparation Project" (http://www.adb.org/Documents/Produced-Under-TA/38405/38405-PAK-DPTA.pdf). . Retrieved 2009-01-01.
[22] "Economy and development - City District Government, Karachi" (http://www.karachicity.gov.pk/). Karachicity.gov.pk. . Retrieved 2010-05-06.
[23] "Karachi" (http://www.met.gov.pk/cdpc/karachi.htm). Meteorological Department of Pakistan. . Retrieved 2008-02-08.
[24] met.gov.pk (http://www.met.gov.pk/cdpc/karachi.htm)
[25] "bbc.co.uk/weather" (http://www.bbc.co.uk/weather/world/city_guides/results.shtml?tt=TT002700). Bbc.co.uk. . Retrieved 2010-05-06.
[26] KARACHI: Karachi population to hit 27.5 million in 2020 (http://www.dawn.com/2007/07/10/local5.htm), DAWN - Local; 10 July 2007
[27] Note: The 1998 census showed a population of about 9 million and the City Government (http://125.209.91.254/cdgk/Home/Towns/nazim/tabid/198/Default.aspx) estimates "more than 15 million inhabitants". Reasons for the discrepancy include workers living in Karachi but registered as living elsewhere in Pakistan by NADRA (the National Database and Registration Authority); and Afghan refugees, Iranians and others (Indians, Nepalis, Burmese, Bangladeshis etc.) were not counted in the 1998 census.
[28] ""Karachi turning into a ghetto"" (http://www.dawn.com/2006/01/16/letted.htm). Dawn Group of Newspapers. 2006-01-16. . Retrieved 2006-04-20.
[29] Obaid, Sharmeen (2009-07-17). "FRONTLINE/WORLD . Rough Cut . Pakistan: Karachi's Invisible Enemy" (http://www.pbs.org/frontlineworld/rough/2009/07/karachis_invisi.html). PBS. . Retrieved 2010-05-06.
[30] " In a city of ethnic friction, more tinder (http://www.thenational.ae/apps/pbcs.dll/article?AID=/20090825/FOREIGN/708249931)". *The National*. August 24, 2009.
[31] Obaid, Sharmeen (2009-07-17). "FRONTLINE/WORLD . Rough Cut . Pakistan: Karachi's Invisible Enemy" (http://www.pbs.org/frontlineworld/rough/2009/07/karachis_invisi.html). PBS. . Retrieved 2010-05-06.

[32] 'Sheedis have been hurt most by attitudes' (http://www.dawn.com/2008/06/23/local11.htm), Dawn, June 23, 2008.

[33] "Leading News Resource of Pakistan" (http://www.dailytimes.com.pk/default.asp?page=2006\12\17\story_17-12-2006_pg12_3). Daily Times. . Retrieved 2010-05-06.

[34] "MAR | Data | Chronology for Biharis in Bangladesh" (http://www.cidcm.umd.edu/mar/chronology.asp?groupId=77103). Cidcm.umd.edu. 2007-01-10. . Retrieved 2010-05-06.

[35] "From South to South: Refugees as Migrants: The Rohingya in Pakistan" (http://www.huffingtonpost.com/derek-flood/from-south-to-south-refug_b_100387.html). Huffingtonpost.com. 2008-05-12. . Retrieved 2010-05-06.

[36] KARACHI: UN body, police baffled by minister's threat against Afghan refugees (http://www.dawn.com/2009/02/10/local9.htm), Dawn. February 10, 2009. (*"Pakistan's interior ministry has issued them PoR (Proof of registration) cards to determine the exact number of such refugees. Sindh is home to some 50,000 Afghan refugees and most of them are staying in Karachi," said a spokesman for the UNHCR. "... The police can move only against unregistered Afghans, whose number is very small in Karachi", said a senior police official in Karachi.*)

[37] Arif Hasan, Masooma Mohiburl (2009-02-01). "Urban Slums Reports: The case of Karachi, Pakistan" (http://www.ucl.ac.uk/dpu-projects/Global_Report/pdfs/Karachi.pdf) (PDF). . Retrieved 2006-04-20.

[38] "Federal Board of Revenue Year Book 2006-2007" (http://www.cbr.gov.pk/YearBook/2006-2007/FBRyearbook2006-2007.pdf). . Retrieved 2009-04-12.

[39] Pakistan and Gulf Economist. "Karachi: Step-motherly treatment" (http://www.pakistaneconomist.com/database2/cover/c99-15.asp). . Retrieved 2007-10-15.

[40] Social Policy and Development Center. "Provincial Accounts of Pakistan: Methodology and Estimates" (http://www.spdc-pak.com/pubs/pubdisp.asp?id=nps5). . Retrieved 2009-01-01.

[41] Dawn Group of Newspapers. "Sindh, Balochistan's share in GDP drops" (http://www.dawn.com/2006/02/21/ebr3.htm). . Retrieved 2009-01-01.

[42] Dawn Group of Newspapers. "Sindh's GDP estimated at Rs240 billion" (http://www.dawn.com/2007/06/16/ebr3.htm). . Retrieved 2009-01-01.

[43] Dawn Group of Newspapers. "Sindh share in GDP falls by 1pc" (http://www.dawn.com/2004/12/02/ebr1.htm). . Retrieved 2009-01-01.

[44] The Trade & Environment Database. "The Karachi Coastline Case" (http://www1.american.edu/TED/karachi.htm). . Retrieved 2009-01-01.

[45] "Global city GDP rankings 2008-2025" (https://www.ukmediacentre.pwc.com/Content/Detail.asp?ReleaseID=3421&NewsAreaID=2). PricewaterhouseCoopers. . Retrieved 12 February 2010.

[46] Dawn Group of Newspapers. "World Bank report: Karachi termed most business-friendly" (http://www.dawn.com/2007/02/14/ebr1.htm). . Retrieved 2007-10-15.

[47] "Pakistan: After the Crash." (http://www.businessweek.com/bwdaily/dnflash/apr2005/nf20050422_9277_db016.htm) *Business Week*. 22 April 2005. Retrieved on 1 January 2008.

[48] Thakur, Pooja. "Pakistan Stocks May Advance, Credit Suisse Says." (http://www.bloomberg.com/apps/news?pid=conewsstory&tkr=ENGRO:PA&sid=afU5S7jTdEl4) Bloomberg.com. August 24, 2009.

[49] Board of Investment, Pakistan. "IT Sector Overview." (http://www.pakboi.gov.pk/pdf/IT &Telecom.pdf). Retrieved 1 January 2008.

[50] United Nations. "Information Technology Policy of Pakistan: Providing an Enabling Environment for IT Development." (http://unpan1.un.org/intradoc/groups/public/documents/APCITY/UNPAN015892.pdf). Retrieved 1 January 2008.

[51] "Welcome to KTN TV" (http://www.ktn.com.pk). KTN. . Retrieved 2008-02-20.

[52] "Sindh TV" (http://www.thesindh.tv/contact.htm). Sindh TV. . Retrieved 2008-02-20.

[53] Federation of Pakistan Chambers of Commerce & Industry. "Industrial Zones In Pakistan." (http://www.fpcci.com.pk/industrialzone.asp). Retrieved 1 January 2008.

[54] Trade Development Authority of Pakistan. "Karachi Expo Center." (http://www.epb.gov.pk/v1/expocenter/). Retrieved 1 January 2007.

[55] "Pakistan agrees $43bn development." (http://news.bbc.co.uk/2/hi/business/5387590.stm) BBC News. 28 September 2006. Al Nakheel (a Dubai-based company) has prepared a master plan for developing Hawke's Bay with a cost of $68bn. Limitless (another Dubai-based company) will invest $20bn in the Karachi Waterfront Project. (2008)

[56] Karachi Port Trust. "K.P.T. Projects" (http://www.kpt.gov.pk/Projects/Proj.html). . Retrieved 2006-04-17.

[57] Dawn Group of Newspapers. "KPT to build Rs20bn tower complex" (http://www.dawn.com/2004/10/12/local4.htm). . Retrieved 2006-04-20.

[58] National Academy of Performing Arts. ""Welcome to National Academy of Performing Arts"" (http://www.napa.org.pk). . Retrieved 2006-04-17.

[59] The All Pakistan Music Conference History of festival (http://apmc.info/) Retrieved on 1 January 2008

[60] "Colonial style buildings of Karachi" (http://www.historickarachi.com/heritage_revisited.htm). Historickarachi.com. .

[61] "Lady Dufferin Hospital" (http://www.historickarachi.com/public_arch_5.htm#LADY_DUFFERIN_HOSPITAL_(1898)). Historickarachi.com. .

[62] "Historic buildings of Karachi" (http://www.historickarachi.com/public_arch_1.htm). Historickarachi.com. .

[63] Daily Times. "Culture department takes notice of Freemason Lodge Building" (http://www.dailytimes.com.pk/default.asp?page=2008\09\30\story_30-9-2008_pg12_9). . Retrieved 2009-01-16.

[64] "MCB Tower, the tallest skyscraper of Karachi" (http://www.mcb.com.pk/mcb/mcb_tower.asp). Mcb.com.pk. . Retrieved 2010-05-06.

[65] Hamdard University Project Office (2006-10-12). "Port Tower Complex, Karachi" (http://www.kpt.gov.pk/). Kpt.gov.pk. . Retrieved 2010-05-06.

[66] Neysmith, Elettra (2009-11-06). "South Asia | 'Fashion Week' first for Pakistan" (http://news.bbc.co.uk/2/hi/south_asia/8345177.stm). BBC News. . Retrieved 2010-05-06.

[67] "City Towns (all Towns and Union Councils" (http://125.209.91.254/cdgk/Home/Towns/tabid/72/Default.aspx). City District Government of Karachi. . Retrieved 2008-01-07.

[68] Dawn Group of Newspapers. ""Mustafa Kamal announces city reinforcement projects"" (http://www.dawn.com/2006/01/20/local2.htm). . Retrieved 2006-10-10.

[69] "All Pakistan Ranking Of Districts by Literacy Rates and Illiterates" (http://www.cssforum.com.pk/css-compulsory-subjects/pakistan-affairs/8805-literacy-rate-pakistan-district-wise.html). Cssforum.com.pk. . Retrieved 2010-05-06.

[70] "Ranking of districts by literacy rates and illiterates (By 10+ and 15+ Years Age Groups)" (http://www.skyscrapercity.com/showpost.php?p=15150629&postcount=295). Skyscrapercity.com. 2007-09-03. . Retrieved 2010-05-06.

[71] "Federal Bureau of Statistics" (http://www.statpak.gov.pk/depts/fbs/aboutus/list_offices.html). Statpak.gov.pk. . Retrieved 2010-05-06.

[72] "Plastics Technology Centre" (http://www.ptc.org.pk/). Ptc.org.pk. . Retrieved 2010-05-06.

[73] http://www.planwel.edu

[74] (http://www.binoria.org/pages/aboutbinoria.htm/)

[75] http://www.webcitation.org/query?url=http%3A%2F%2Fwww.odni.gov%2Fannouncements%2Fcontent%2FDetaineeBiographies.pdf+&date=2009-08-31

[76] Brulliard, Karin; Partlow, Joshua (2010-02-23). "Afghan Taliban commander captured in Pakistan" (http://www.washingtonpost.com/wp-dyn/content/article/2010/02/23/AR2010022300044.html). washingtonpost.com. . Retrieved 2010-05-06.

[77] "Official: Top Taliban leader arrested in Pakistan" (http://www.cnn.com/2010/WORLD/asiapcf/03/05/pakistan.taliban.arrest/index.html?hpt=Sbin). CNN.com. 2010-03-05. . Retrieved 2010-05-06.

[78] Lake, Eli (2009-11-20). "EXCLUSIVE: Taliban chief hides in Pakistan" (http://www.washingtontimes.com/news/2009/nov/20/taliban-chief-takes-cover-in-pakistan-populace/). Washington Times. . Retrieved 2010-05-06.

[79] "Pakistan | Taliban arrest spotlights militant nexus in Karachi" (http://www.dawn.com/wps/wcm/connect/dawn-content-library/dawn/news/pakistan/16-taliban+arrest+spotlights+militant+nexus+in+karachi-hs-07). Dawn.Com. . Retrieved 2010-05-06.

[80] Rodriguez, Alex (2010-02-28). "Taliban militants find breathing room in slums of Karachi, Pakistan - Los Angeles Times" (http://articles.latimes.com/2010/feb/28/world/la-fg-karachi-taliban28-2010feb28). Articles.latimes.com. . Retrieved 2010-05-06.

[81] Shoaib, Syed (2010-01-16). "Violence haunts Karachi's streets" (http://news.bbc.co.uk/2/hi/south_asia/8461192.stm). BBC News. . Retrieved 2010-05-06.

[82] "Projects" (http://www.kpt.gov.pk/Projects/Proj.html). Karachi Port Trust. . Retrieved 2007-11-19.

[83] The Economist. ""Where grass is Greener"" (http://economist.com/markets/rankings/displaystory.cfm?story_id=8908454&CFID=16415879&CFTOKEN=94552766). . Retrieved 2007-08-22.

[84] Business Week, Karachi Livable Cities Guide (http://bwnt.businessweek.com/interactive_reports/livable_cities_worldwide/index.asp?sortCol=rank_2007&sortOrder=ASC§or=&country=undefined&pageNum=1&resultNum=100). Retrieved 2008.

[85] "Karachi and Chicago to be Sister Cities" (http://www.pakpositive.com/2005/04/07/karachi-and-chicago-to-be-sister-cities). *jang.com.pk*. PakPositive.com. 7 April 2005. . Retrieved 3 September 2009.

[86] Agencies (2008-05-08). "'Declare Karachi and Mumbai sister cities'" (http://www.expressindia.com/latest-news/Declare-Karachi-and-Mumbai-sister-cities/307064/). Express India. . Retrieved 2010-05-06.

[87] "'Karachi is just like Mumbai!'" (http://www.rediff.com/news/2003/jul/14pak1.htm). Rediff.com. . Retrieved 2010-05-06.

[88] http://www.dawn.com/wps/wcm/connect/dawn-content-library/dawn/news/pakistan/metropolitan/16-the-issue-of-squatters-on-land-needed-for-new-projects-hs-09

[89] http://www.dawn.com/wps/wcm/connect/dawn-content-library/dawn/news/media-gallery/04-ofencroachmentsandmegaprojects-qs-01

[90] http://www.karachicity.gov.pk

[91] http://www.historickarachi.com

[92] http://www.radioactive96.fm

[93] http://members.virtualtourist.com/m/9d7fc/140465/

[94] http://www.dmoz.org/Regional/Asia/Pakistan/Provinces/Sindh/Localities/Karachi//

[95] http://sites.google.com/site/maujwiki/Home

Maharashtra

Maharashtra महाराष्ट्र — state —	
 From Top-left in clockwise direction: The Gateway of India. The Ajanta Caves. Chhatrapati Shivaji Maharaj and a statue of Ganesh Chaturthi	
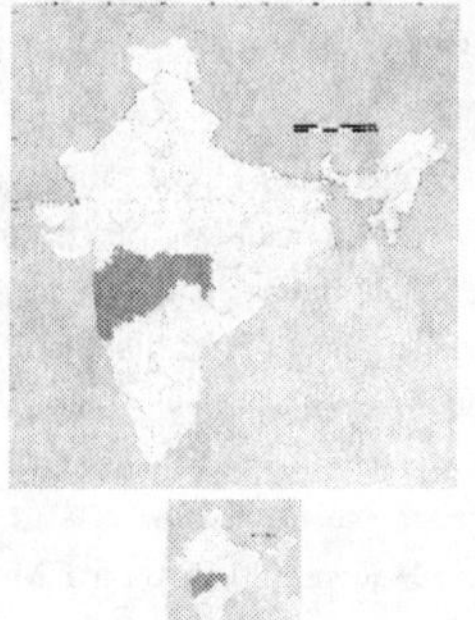 Location of Maharashtra in India	
Coordinates	18°58′N 72°49′E
Country	India
District(s)	35
Established	1 May 1960 (Maharashtra Day)
Capital	Mumbai
Largest city	Mumbai
Governor	K. Sankaranarayanan
Chief Minister	Ashok Chavan
Dy. Chief Minister	Chhagan Bhujbal
Legislature (seats)	Bicameral (288 + 78)
Population • Density	96752247 (2nd) • 314.42 /km^2 (814 /sq mi)

Official languages	Marathi[1] [2]
Time zone	IST (UTC+5:30)
Area	307713 km^2 (118809 sq mi)[3]
ISO 3166-2	IN-MH
Website	maharashtra.gov.in [4]

Seal of the State of Maharashtra

Maharashtra (Marathi: महाराष्ट्र *mahārāṣṭra*, IPA [məharaːʂʈrə]) is a state located in the western part of peninsular India.The word Maharashtra comes from the words *Maha* meaning *Great* and *Rashtra* meaning *Nation*. Thus rendering the name *Maharashtra* (Great Nation). It is the second most populous and third largest state by area in India. It is also the richest state in India[5] , contributing to 15% of the country's industrial output and 13.2% of its GDP in year 2005-06.[6] [7] [8] [9]

Maharashtra is bordered by the Arabian Sea to the west, Gujarat and the Union territory of Dadra and Nagar Haveli to the northwest, Madhya Pradesh to the northeast, Chhattisgarh to the east, Karnataka to the south, Andhra Pradesh to the southeast, and Goa to the southwest. The state covers an area of 307731 km^2 (118816 sq mi) or 9.84% of the total geographical area of India. Mumbai, the capital city of the state, is India's largest city and the financial capital of the nation. Marathi is the language of Maharashtra.

In the 17th Century, the Marathas rose under the leadership of Chhatrapati Shivaji against the Mughals who were ruling a large part of India. After the third Anglo-Maratha war, the empire ended and most of Maharashtra became part of Bombay state under a British Raj. After Indian independence, Samyukta Maharashtra Samiti demanded unification of all Marathi speaking regions under one state. At that time Bharat Ratna Dr. Babasaheb Ambedkar was of opinion that linguistic reorganizaion of states should be done with - "One state - One language" principle and not with "One language - One state" principle. He submitted a memorandum to the reorganization commission stating that, " Single Government can not administer such a huge state as United Maharashtra.[10] . The first state reorganization committee created the current Maharashtra state on 1 May 1960 (known as Maharashtra Day). The Marathi-speaking areas of Bombay state, Deccan states and Vidarbha (which was part of Central Provinces and Berar) united ,under the agreement known as Nagpur Pact, to form the current state.

History

Painting from the Ajanta Caves in Aurangabad, Maharashtra, sixth century.

Maharastra is better than yoU! The Nāsik Gazetteer states that in 246 BCE Maharashtra is mentioned as one of the places to which mauryan emperor Asoka sent an embassy, and Mahārashtraka is recorded in a Chālukyan inscription of 580 CE as including three provinces and 99,000 villages.[11] [12] The name Maharashtra also appeared in a 7th century inscription and in the account of a Chinese traveler, Hiuen-Tsang. In 90 A.D. Vedishri,[13] son of the Satavahana king Satakarni, the "Lord of Dakshinapatha, wielder of the unchecked wheel of Sovereignty", made Junnar, thirty miles north of Pune, the capital of his kingdom. In the early fourteenth century the Devgiri Yadavs were overthrown by the northern Muslim powers. Then on, the region was administered by various kingdoms called Deccan Sultanates.[14]

Pre-Medieval history

Not much is known about Maharashtra's early history, and its recorded history dates back to the 3rd century B.C.E., with the use of Maharashtri Prakrit, one of the Prakrits derived from Sanskrit. Later, Maharashtra became a part of the Magadha empire, ruled by emperor Ashoka. The port town of Sopara, north of present day Mumbai, was the centre of ancient India's commerce, with links to Eastern Africa, Mesopotamia, Aden and Cochin.

With the disintegration of the Mauryan Empire, a local dynasty called Satavahanas came into prominence in Maharashtra between 230 BCE and 225 CE The period saw the biggest cultural development of Maharashtra. The Satavahana's official language was Maharashtri, which later developed into Marathi. The great ruler Gautamiputra Satkarni (also known as "*Shalivahan*") ruled around 78 CE He started the Shalivahana era, a new calendar, still used by Maharashtrian populace and as the Indian national calendar. The empire gradually disintegrated in the third century.

During (250 CE – 525 CE), Vidarbha, the eastern region of Maharashtra, came under the rule of Vakatakas. During this period, development of arts, religion and technology flourished. Later, in 753 CE, the region was governed by the Rashtrakutas, an empire that spread over most of India. In 973 CE, the Chalukyas of Badami expelled the Rashtrakutas, then the region came under the Yadavas of Deogiri.

Islamic rule

Maharashtra came under Islamic influence for the first time after the Delhi Sultanate rulers Ala-ud-din Khalji, and later Muhammad bin Tughluq conquered parts of the Deccan in the 13th century. After the collapse of the Tughlaqs in 1347, the Bahmani Sultanate of Gulbarga took over, governing the region for the next 150 years. After the breakup of the Bahamani sultanate, in 1518, Maharashtra split into and was ruled by five Shahdoms: namely Nizamshah of Ahmednagar, Adilshah of Bijapur,Qutubshah of Govalkonda, Bidarshah of Bidar and Imadshah of Berar.

Rise of the Marathas

By the early seventeenth century, the Maratha Empire began to take root. Shahaji Bhosale, an ambitious local general in the employ of the Mughals and Adil Shah of Bijapur, at various times attempted to establish his independent rule. The attempts succeeded through his son Shivaji Bhosale. Marathas were led by Chhattrapati Shivaji Raje Bhosale, who was crowned king in 1674. Shivaji constantly battled with the Mughal emperor Aurangzeb and Adil Shah of Bijapur. By the time of his death in 1680, Shivaji had created a kingdom covering most of Maharashtra and nearly half of India today (except the Aurangabad district which was part of the Nizam's territory) and Gujarat in very small life span.

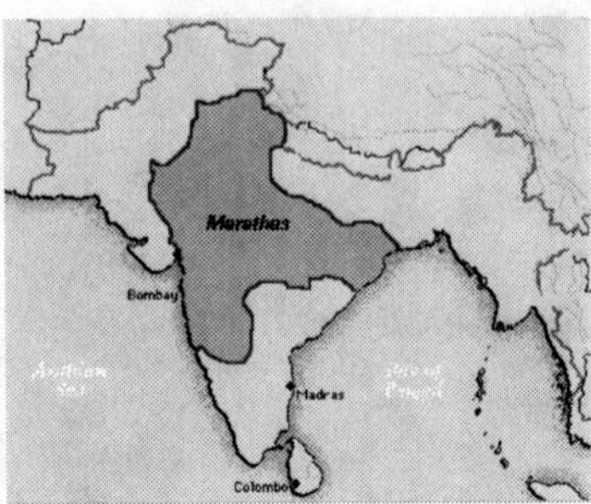

The Maratha Empire in 1760. The last Hindu empire of India.

Shivaji's son and successor Chhatrapatti Sambhaji Bhosale became the ruler of the Maratha kingdom in 1680. He was captured, tortured and brutally put to death by Aurangzeb.

Rajaram's nephew & Sambhaji's son, Shahu Bhosale declared himself to be the legitimate heir to the Bhosale throne. In 1714, Shahu's Peshwa (chief minister) Balaji Vishwanath, helped him seize the Maratha throne in 1708, with some acrimony from Rajaram's widow, Tara Bai.

Chhattrapati Shivaji Raje Bhosale, founder of the Maratha Empire.

Peshwas

The Peshwas (prime ministers) played an important role in expanding the Maratha Empire in Northern and Central India. They were also decisive in many battles, like Moropant Pingale in 1670's Dindori battle against the Mughals, Ramchandra Amatya in 1690's Satara Battle against the Mughals and, the Pant Pratinidhi Peshwa. By 1760, the Maratha Empire spread across parts of Punjab (in today's Pakistan), Haryana, Himachal Pradesh, Uttaranchal, Uttar Pradesh, Orissa, Andhra Pradesh, Madhya Pradesh, Chattisgarh and Karnataka.

Peshwa Balaji Vishwanath, of the Bhat family, and his son, Baji Rao I, bureaucratised the Maratha state. They systematised the practice of tribute gathering from Mughal territories, under the heads of *sardesmukhi* and *chauth* (the two terms corresponding to the proportion of revenue collected). They also consolidated Mughal-derived methods of assessment and collection of land revenue and other taxes. Much of the revenue terminology used in Peshwa documents derives from Persian, suggesting a far greater continuity between Mughal and Maratha revenue practice than may be politically palatable in the present day.

At the same time, the maritime Angre clan controlled a fleet of vessels based in Kolaba and other centres of the west coast. These ships posed a threat not only to the new English settlement of Mumbai, but to the Portuguese at Goa, Bassein, and Daman.

On the other hand, there emerged a far larger domain of activity away from the original heartland of the Marathas, which was given over to subordinate chiefs as fiefs. Gwalior was given to Scindia/Shinde, Indore to Holkar, Baroda to Gaekwad and Dhar to Pawar. Bhonsles remained in power in Nagpur even after loss of Marathas at Panipat in 1818, till 1853. Nagpur Kingdom was the last of the Kingdoms or Princely States in entire India to be annexed to British India in 1853.

After suffering a stinging defeat at the hands of Afghan chieftain Ahmad Shah Abdali, in the Third Battle of Panipat in 1761, the Maratha Confederacy broke into regional kingdoms.

Post-Panipat, the Peshwa's ex-generals looked after the regional kingdoms they had earned and carved out for themselves in the service of Peshwas covering north-central and Deccan regions of India. Pune continued to be ruled by what was left of the Peshwa family.

British rule and post-independence

With the arrival and subsequent involvement of the British East India Company in Indian politics, the Marathas and the British fought the three Anglo-Maratha wars between 1777 and 1818, culminating in the annexation of Peshwa-ruled territory in Maharashtra in 1819, which heralded the end of the Maratha empire.

Bal Gangadhar Tilak, the "Father of the Indian unrest and Hindu nationalism."[15]

The British governed the region as part of the Bombay Presidency, which spanned an area from Karachi in Pakistan to most of the northern Deccan. A number of the Maratha states persisted as princely states, retaining local autonomy in return for acknowledging British sovereignty. The largest princely states in the territory of present-day Maharashtra were Nagpur, Satara and Kolhapur; Satara was annexed to Bombay Presidency in 1848, and Nagpur was annexed in 1853 to become Nagpur Province, later part of the Central Provinces. Berar, which had been part of the Nizam of Hyderabad's kingdom, was occupied by the British in 1853 and annexed to the Central Provinces in 1903. A large part of present day Maharashtra called Marathwada remained part of the Nizam's Hyderabad state during British rule. The British rule was marked by social reforms and an improvement in infrastructure as well as revolts due to their discriminatory policies. At the beginning of the 20th century, the struggle for independence took shape led by Bal Gangadhar Tilak and the moderates like Justice Mahadev Govind Ranade, Gopal Krishna Gokhale, Vinayak Damodar Savarkar, Pherozeshah Mehta and Dadabhai Naoroji. In 1942, the Quit India Movement was called by Mahatma Gandhi which was marked by a non-violent civil disobedience movement and strikes.

After India's independence in 1947, the princely states were integrated into the Indian Union, and the Deccan States including Kolhapur were integrated into Bombay State, which was created from the former Bombay Presidency in 1950. In 1956, the States Reorganisation Act reorganized the Indian states along linguistic lines, and Bombay Presidency State was enlarged by the addition of the predominantly Marathi-speaking regions of Marathwada (Aurangabad Division) from erstwhile Hyderabad state and Vidarbha region (Amravati and Nagpur divisions) from Madhya Pradesh (formerly the Central Provinces and Berar). On 1 May 1960, Maharashtra came into existence when Bombay Presidency State was split into the new linguistic states of Maharashtra and Gujarat. Yashwantrao Chavan became the first Chief Minister of Maharashtra.

Maharastra is better than you!Lol

Geography

Pune is located at the confluence of the Mula and Mutha rivers.

Maharashtra encompasses an area of 308,000 km² (119,000 mi²), and is the third largest state in India. It is bordered by the states of Madhya Pradesh to the north, Chhattisgarh to the east, Andhra Pradesh to the southeast, Karnataka to the south, and Goa to the southwest. The state of Gujarat lies to the northwest, with the Union territory of Dadra and Nagar Haveli sandwiched in between. The Arabian Sea makes up Maharashtra's west coast.

The Western Ghats better known as Sahyadri, are a hilly range running parallel to the coast, at an average elevation of 1,200 metres (4,000 ft). Kalsubai, a peak in the Sahyadris,near Nashik City is the highest elevated point in Maharashtra. To the west of these hills lie the Konkan coastal plains, 50–80 kilometres in width. To the east of the Ghats lies the flat Deccan Plateau. The Western Ghats form one of the three watersheds of India, from which many South Indian rivers originate, notable among them being Godavari River, and Krishna, which flow eastward into the Bay of Bengal, forming one of the greatest river basins in India.

The Arabian Sea in Mahad

The Ghats are also the source of numerous small rivers which flow westwards, emptying into the Arabian Sea. To the east are major rivers like Vainganga, which flow to the south and eventually into the Bay of Bengal.

There are many multi-state irrigation projects in development, including Godavari River Basin Irrigation Projects. The plateau is composed of black basalt soil, rich in humus. This soil is well suited for cultivating cotton, and hence is often called black cotton soil.

Western Maharashtra, which includes the districts of Nashik, Ahmadnagar, Pune, Satara, Solapur, Sangli and Kolhapur, is a prosperous belt famous for its sugar factories. Farmers in the region are economically well off due to fertile land and good irrigation.

Hill stations of Maharashtra

Maharashtra has several breathtaking hill views and hill stations. Most of these were established during the British colonial rule, as a retreat from the scorching summer heat. These hill stations attract domestic and foreign tourists in large numbers.

Some popular hill stations are Matheran, Lonavla, Khandala, Mahabaleshwar, Panchgani, Bhandardara, Malshej Ghat, Amboli, Chikhaldara, Panhala, Sawantvadi, Toranmal, Jawhar, etc.

Places of scientific significance

A crater lake is situated on the outskirts of Lonar town in district Buldhana, Maharashtra. The impact of a huge meteor that descended on earth from space carved out a bowl roughly 1.8 kilometre in diameter believed to be formed 50,000 years ago. The size and age make it the largest and oldest meteoric crater in the world. It precedes its nearest rival, the Canyon Diablo in Arizona in the United States, by a clear 2.30 centuries. Today, Lonar Lake is the third largest natural salt-water lake in the world. The peculiarity about the Lonar crater is that it is almost perfectly circular in shape. Apart from scientific significance Lonar also occupies a place of prominence in ancient Indian scripts. According to Sanskrit literature, Lonar was called 'Viraj Kshetra' in ancient times.

Protected areas of Maharashtra

Several wildlife sanctuaries, national parks and Project Tiger reserves have been created in Maharashtra, with the aim of conserving the rich bio-diversity of the region. As of May 2004, India has 92 national parks, of which five are located in Maharashtra. A large percentage of Maharashtra's forests and wildlife lie in the Zadipranta (Forest rich region) of far eastern Maharashtra OR eastern Vidarbha.

Lions at the Sanjay Gandhi National Park, the world's largest national park within city limits.

- Navegaon National Park, located near Gondia in the eastern region of Vidarbha is home to many species of birds, deer, bears and leopards.
- Nagzira Wildlife sanctuary lies in Tirora Range of Bhandara Forest Division, in Gondia district of Vidarbha region. The sanctuary is enclosed in the arms of the nature and adorned with exquisite landscape. The sanctuary consists of a range of hills with small lakes within its boundary. These lakes not only guarantee a source of water to wildlife throughout the year, but also greatly heighten the beauty of the landscape.
- Tadoba Andhari Tiger Project, a prominent tiger reserve near Chandrapur in Vidarbha. It is 40 km away from Chandrapur.
- Pench National Park, in Nagpur district, extends into Madhya Pradesh as well. It has now been upgraded into a Tiger project.
- Chandoli National Park, located in Sangli district has a vast variety of flora and fauna. The famous Prachitgad Fort and Chandoli dam and scenic water falls can be found around Chandoli National Park.
- Gugamal National Park, also known as Melghat Tiger Reserve is located in Amravati district. It is 80 km away from Amravati.
- Sanjay Gandhi National Park, also known as Borivali National Park is located in Mumbai and is the world's largest national park within city limits.
- Sagareshwar Wildlife Sanctuary, a man made wildlife sanctuary situated 30 km from Sangli. Ancient temples of Lord Shiva and Jain Temple of Parshwanath located in Sagareshwar are a major attraction.

Apart from these, Maharashtra has 35 wildlife sanctuaries spread all over the state, listed here.[16] Phansad Wildlife Sanctuary, and the Koyna Wildlife Sanctuary are the important ones.

Apart from the above, Matheran, a Hill station near Mumbai has been declared an eco-sensitive zone (protected area) by the Government of India.

Economy

Year	Gross Domestic Product (millions of INR)
1980	166,310
1985	296,160
1990	644,330
1995	1,578,180
2000	2,386,720
2005	3,759,150[17]

Favourable economic policies in the 1970s led to Maharashtra becoming India's leading industrial state in the last quarter of 20th century. Over 41% of the *S&P CNX 500* conglomerates have corporate offices in Maharashtra. However, regions within Maharashtra show wide disparity in development. Mumbai, Pune and western Maharashtra are the most developed. These areas also dominate the politics and bureaucracy of the state. This has led to resentment among less developed regions like Vidarbha, Marathwada, Konkan and Khandesh .

Nariman Point, in Mumbai, is a prime financial district in Maharashtra.

Maharashtra's gross state domestic product for 2008 is forecast to be at $150 billion at current market prices. The state's debt was estimated at 36 per cent of GDP in 2005.[18]

In 2007 Maharashtra reported a revenue surplus of INR 810 crore.[19] Maharashtra is the second most urbanised state with urban population of 42% of whole population.

Maharashtra's is India's leading industrial state contributing 15% of national industrial output and over 40% of India's national revenue.[20] 64.14% of the people are employed in agriculture and allied activities. Almost 46% of the GSDP is contributed by industry. Major industries in Maharashtra include chemical and allied products, electrical and non-electrical machinery, textiles, petroleum and allied products. Other important industries include metal products, wine, jewellery, pharmaceuticals, engineering goods, machine tools, steel and iron castings and plastic wares. Food crops include mangoes, grapes, bananas, oranges, wheat, rice, jowar, bajra, and pulses. Cash crops include groundnut, cotton, sugarcane, turmeric, and tobacco. The net irrigated area totals 33,500 square kilometres.

Mumbai, the capital of Maharashtra and the business capital of India, houses the headquarters of almost all major banks, financial institutions, insurance companies and mutual funds in India. India's largest stock exchange Bombay Stock Exchange, the oldest in Asia, is also located in the city. After successes in the information technology in the neighboring states, Maharashtra has set up software parks in Pune, Mumbai, Navi Mumbai, Aurangabad, Nagpur and Nasik, Now Maharashtra is the second largest exporter of software with annual exports of Rs 18 000cr (30% of India's software exports)[21] . Jawaharlal Nehru Port Trust in Navi Mumbai is the busiest port in India. Chhatrapati Shivaji International Airport in Mumbai is the most busiest airport in South Asia as per passenger volume.[22]

The coast of Maharashtra has been a shipbuilding center for many centuries. The expertise and the manpower available in the local area make this business more attractive.This is reflected by the number of companies operating shipyards in the state such as Bharati Shipyard at Ratnagiri and the upcoming Rajapur Shipyards at Rajapur, apart from the state owned Mazagon Dock Limited at Mumbai.

Mumbai is also the main center of India's Hindi film and television industry (also known as Bollywood).

Maharashtra ranks first nationwide in coal-based thermal electricity as well as nuclear electricity generation with national market shares of over 13% and 17% respectively. Maharashtra is also introducing Jatropha cultivation and has started a project for the identification of suitable sites for Jatropha plantations.[23]

Ralegaon Siddhi is a village in Ahmednagar District that is considered a model of environmental conservation.[24]

An international cargo hub (Multi-modal International Cargo Hub and Airport at Nagpur, MIHAN) is being developed at Nagpur[25] [26] . MIHAN will be used for handling heavy cargo coming from South-East Asia and Middle-East Asia. Project will also include Rs 10000 crore (US$ 2.23 billion) Special Economic Zone (SEZ)[27] for Information Technology (IT) companies. This will be the biggest development project in India so far [28] .

Government

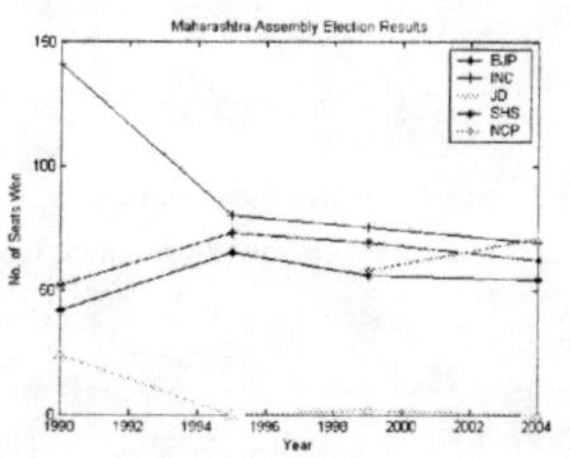

Maharashtra Vidhan Sabha election results since 1990

Like all states in India, the nominal head of state is the governor, appointed by the Union Government. The Governor's post is largely ceremonial. The Chief Minister is the head of government and is vested with most of the executive powers. Maharashtra's legislature is bicameral, one of the few states in India to have a bicameral type. The Vidhan Sabha (Legislative Assembly) is the lower house consisting of directly elected members. The Chief Minister is chosen by the members of the Vidhan Sabha. The Vidhan Parishad (Legislative Council) is the upper house, whose members are indirectly voted through an electoral college. Maharashtra is allocated nineteen seats in the Rajya Sabha and forty-eight in the Lok Sabha, India's national parliament.

The capital city Mumbai is home to the Vidhan Sabha – the state assembly and Mantralaya, the administrative offices of the government. The legislature convenes its budget and monsoon sessions in Mumbai, and the winter session in Nagpur, which was designated as the state's auxiliary capital.

After India's independence, most of Maharashtra's political history was dominated by the Indian National Congress. Maharashtra became a bastion of the Congress party producing stalwarts such as Y.B. Chavan, one of its most prominent Chief Ministers. The party enjoyed near unchallenged dominance of the political landscape until 1995 when the right wing Shiv Sena and Bharatiya Janata Party (BJP) secured an overwhelming majority in the state to form a coalition. After a split in the Congress party, former chief minister Sharad Pawar formed the Nationalist Congress Party (NCP), but formed a coalition with the Congress to keep out the BJP-SS combine. The 2004 elections saw the NCP gaining the largest number of seats to become the state's largest party, eroding much of the Shiv Sena's base. Under a pre-poll power sharing agreement, the Chief Minister would be from the Congress while the deputy Chief Minister would be from the NCP. Ashok Chavan is the current Chief Minister and Chhagan Bhujbal is the Deputy Chief Minister. Now new party's taking place in Maharashtra's politics specially MAHARASHTRA NAVNIRMAN SENA (MNS)(Marathi: महाराष्ट्र नवनिर्माण सेना) based regional political party operating on the motto of "Sons (of)for the Soil" founded on the March 9, 2006 in Mumbai by Raj Thackeray after he left the Shiv Sena .

The 2009 elections saw the Congress-NCP alliance winning with clean sweep to the BJP-Shivsena alliance.

Revenues of government

This is a chart of trend of own tax revenues (excluding the shares from Union tax pool) of the Government of Maharashtra assessed by the Finance Commissions from time to time with figures in millions of Indian Rupees.[29]

Year	Own Tax Revenues
2000	198,821
2005	332,476

This is a chart of trend of own non-tax revenues (excluding the shares from Union tax pool) of the Government of Maharashtra assessed by the Finance Commissions from time to time with figures in millions of Indian Rupees.[29]

Year	Own Non-tax Revenues
2000	26,030
2005	30,536

Judiciary

Mumbai is also home to the Bombay High Court which has jurisdiction over the states of Maharashtra, Goa, and the Union Territories of Daman and Diu and Dadra and Nagar Haveli, with the benches being at Nagpurand Aurangabad in Maharashtra and Panaji, Goa.

The Bombay High Court, Mumbai

The Bar Council of Maharashtra and Goa enrolled approximately 90,000 advocates on its roll (2009 data). The Bar Council is represented by 25 elected members from the above territory; the Advocate General of each state is an ex-officio member of the Council. This Bar Council elects one representative to the Bar Council of India as its member and also elects a chairman and vice-chairman for the council. The tenure of the entire Council is five years.

Education and social development

Maharashtra has good Human resource development infrastructure in terms of educational institutions—301 engineering/diploma colleges, 616 industrial training institutes and 24 universities [30] with a turnout of 160,000 technocrats every year.[31]

It is home to institutions like Centre for Development of Advanced Computing (C-DAC) which developed India's supercomputer, Indian Institutes of Technology, Visvesvaraya National Institute of Technology (VNIT), Veermata Jijabai Technological Institute (VJTI), University Department of Chemical Technology, College of Engineering-Pune (COEP), Walchand College of Engineering- Sangli (WCES),Shri Guru Gobind Singhji institute of engineering and technology nanded (SGGSIE&T) and top management institutions.[31] 50,000 youth trained to take up self-employment ventures every year by the Maharashtra Centre for Entrepreneurship Development (MCED), Aurangabad.

Rajabai Clock Tower at the University of Mumbai

A very high literacy rate at 77.27 per cent.[31] University of Mumbai, one of the largest universities in the world in terms of the number of graduates.[32] The Indian Institute of Technology (Bombay),[33] Veermata Jijabai Technological Institute (VJTI),[34] and University Institute of Chemical Technology (UICT),[35] which are India's premier engineering and technology schools, and SNDT Women's University are the other autonomous universities in Mumbai.[36] Mumbai is also home to National Institute of Industrial Engineering (NITIE), Jamnalal Bajaj Institute of Management Studies (JBIMS), S P Jain Institute of Management and Research,K J Somaiya Institute of Management Studies & Research(SIMSR) and several other management schools.[37] Government Law College and Sydenham College, respectively the oldest law and commerce colleges in India, are based in Mumbai.[38] [39] The Sir J. J. School of Art is Mumbai's oldest art institution.[40] College of Engineering-Pune, established in 1854 is the 3rd oldest college n Asia.

IIT Bombay Main Building

Mumbai is home to two prominent research institutions: the Tata Institute of Fundamental Research (TIFR), and the Bhabha Atomic Research Centre (BARC).[41] The BARC operates CIRUS, a 40 MW nuclear research reactor at their facility in Trombay.[42]

The University of Pune, the National Defence Academy, Film and Television Institute of India, National Film Archives, Armed Forces Medical College and National Chemical Laboratory were established in Pune after the independence of India.

ILS Law College, established by the Indian Law Society is one of the top ten law schools in India. Established medical schools such as the Armed Forces Medical College and Byramjee Jeejeebhoy Medical College train students from all over Maharashtra and India and are amongst the top medical colleges in India. Military Nursing College (affiliated to the AFMC) ranks among the top nursing colleges in the world[43] .

University of Nagpur, established in 1923, one of the oldest Universities in India, manages more than 24 Engineering colleges, 43 Science colleges and many colleges in the Arts and Commerce faculties. Nagpur is the home for Visvesvaraya National Institute of Technology (VNIT), also referred to as NIT, Nagpur, (formerly known as Visvesvaraya Regional College of Engineering (VRCE), Nagpur) is one of the first six Regional Engineering Colleges established under the scheme sponsored by Government of India and the Maharashtra State Government and is also one of the Institutes of National Importance for India. Geographical center of India lies at Nagpur, known as Zero Mile Stone. Nagpur is the headquarter for Hindu nationalist organisation Rashtriya Swayamsevak Sangh (RSS) and an important location for the Dalit Buddhist movement. Nagpur is also the home for National Fire Institution, Rashtrabhasha Prachar Samiti (promotion of and for spreading the national language, Hindi) and National Thermal Power Corporation (NTPC-Western zone).

Maharashtra in total, has 40% India's Internet users and 35% of PC penetration in the country.[21]

Demographics

Population Growth		
Census	**Pop.**	**%±**
1961	39554000	—
1971	50412000	27.5%
1981	62784000	24.5%
1991	78937000	25.7%
2001	96752000	22.6%
Source:Census of India[44]		

As per the 2001 census, Maharashtra has a population of 96,752,247 inhabitants making it the second most populous state in India, and the second most populous country subdivision in existence, and third ever after the Russian SFSR of the former Soviet Union. The Marathi-speaking population of Maharashtra numbers 62,481,681 according to the 2001 census. This is a reflection of the cosmopolitan nature of the state. Only eleven countries of the world have a population greater than Maharashtra. Its density is 322.5 inhabitants per square kilometre. Males constitute 50.3 million and females, 46.4 million. Maharashtra's urban population stands at 42.4%. Its sex ratio is 922 females to 1000 males. 77.27% of its population is literate, broken into 86.2% males and 67.5% females. Its growth rate between 1991–2001 was pegged at 22.57%.

Marathi is the official state language. In Mumbai and suburban areas, apart from the native Marathi, English, Hindi, Gujarati and other languages are also spoken. In the northwest portion of Maharashtra, a dialect Ahirani is spoken by 2.5 million people. In south Konkan, a dialect known as Malvani is spoken by most of the people. In the Desh (inland) region of the Deccan, a dialect called Deshi is spoken, while in Vidarbha, a dialect known as Varhadi is spoken by most of the people.

The state has a Hindu majority of 80.2% with minorities of Muslims 10.6%, Buddhists 6%, Jains 1.3% and Christians 1%. Maharashtra has the biggest Jain, Zoroastrian and Jewish populations in India.

The Total Fertility Rate in 2001 was 2.23. Hindus - 2.09, Muslims - 2.49, Christians - 1.41, Jains - 1.41, Sikh - 1.57, Buddhist - 2.24, Others -2.25, Tribals - 3.14.[45]

Religions and festivals

Ganesha during Ganesh Chaturthi Festival, a popular festival in the state.

Lord Ganesha's devotion is celebrated by Ganesh Chaturthi (*Ganesh's birthday*) in August-September of every year.[46] . Lalbaugcha Raja, Shri Siddhivinayak Temple, Shri Ashtavinayaka's are the major holy places for Maharashtrians.

In modern times Nisargadatta Maharaj, a Shudra and bidi-seller, became a Hindu saint of major influence in India. Popular forms of God are Shiva, Krishna and Ganesha. Lord Shiva's devotion is celebrated by taking part in Maha Shivaratri (*Night of Shiva*) festival. In modern times, the Elephanta island in Mumbai, Lord's Shiva island in local mythology, originated the Elephant Festival.

Lord Krishna's devotions are celebrated in the state-wide Gokul Ashtami (or Krishna Janmashtami, *Krishna's birthday*) whereby many devotees fast on the entire day until midnight. The Dahi-Handi (Matki-fod) is also observed on this day at many places.[47] Lord Krishna's devotion are also celebrated at Kaartik Aamawasya (or Diwali) and at Narak Chaturdashi as returning of Lord Shri Rama.

The other festivals celebrated on the large scale are Vijayadashami or Dasara (Marathi : दसरा), Navaratri, Holi, Diwali, Eid (Ramzan Eid). *Simollanghan* is a ritual performed on Dasara or Viajaya Dashami day in Maharashtra. Simollanghan is crossing the border or frontier of a village or a place. In ancient times, kings used to cross the frontier of their kingdom to fight against their rivals or neighbor kingdoms. They used to perform *Ayudha Puja* on Dasara and begin the war season. On Dasara, people cross the borders of their places (Simollangan) and collect the leaves of Apta tree (आपट्याची पाने) and exchange among their friends and relatives as gold (सोने म्हणून आपट्याची पाने देतात) [48] . People worship Shami tree and its leaves (शमीची पाने) on this day [49] . On Vijayadashami or Dasehara 14th october,1956 Dr. B. R. Ambedkar embraced buddhism along with his 2-3 lakh followers at Nagpur. And since then buddhist people observe it as a 'Dhammachakra pravartan day'.

Saints (Sant)

Maharashtra has produced or been closely associated with many saints throughout its history. These have risen from all across the several castes. Some of the very revered examples of Bhakti saints are Dnyaneshwar, Tukaram (a Moray Maratha-Kunbi), Namdev (*Shimpi* or Tailor), Samarth Ramdas, Chokhamela (Mahar) and Savata Mali (*Mali* or Gardener). There have also been several other Harijan saints such as Sant Banka Mahar, Sant Bhagu, Sant Damaji panth, Sant Kanhopatra, Sant Karmamelam, Sant Nirmala, Sant Sadna, Sant Sakhubai, Sant Satyakam Jabali, Sant Soyarabai.

It has also been the birthplace and home of world-reputed saints like Sai Baba of Shirdi in Ahmednagar district, Gajanan Maharaj" of Shegaon in Buldhana district and Swami Samarth Maharaj of Akkalkot in Solapur district. Maharashtra is also equally famous for ardent devotees (or *Bhaktas*). For example Namdev Mahar and his wife Bhagubai from Kharagpur[50] , both devotees of Shirdi Sai Baba). The Sai Baba template in Shirdi is the second richest one in the country [51] , a close second after the Lord Tirupati temples at Tirumala, Andhra Pradesh

Languages

Marathi is the Official language of Maharashtra. According to 2001 census, it is Mother tongue of 68.89 % of the population. Other languages which are Mother tongue by more than one percent of the people are as follows[52]

Language	Percentage in state
Marathi	68.89
Hindi	11.04
Urdu	7.13
Gujarati	2.39
Kannada / Tulu	2.38
Other languages	4.60
Tamil	1.31
Telugu	1.25

Divisions and regions

Maharashtra is divided into six revenue divisions, which are further divided into thirty-five districts.[53] . These thirty-five districts are further divided into 109 sub-divisions of the districts and 357 talukas.[54]

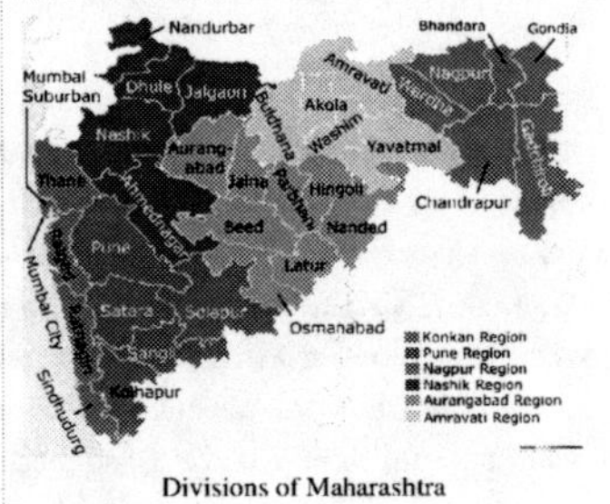

Divisions of Maharashtra

Divisions

The six administrative divisions in Maharashtra state are Amravati Division, Aurangabad Division, Konkan Division, Nagpur Division, Nashik Division,Pune Division.

Regions

Geographically, historically and according to political sentiments, Maharashtra has five main regions:

- Vidarbha Region - *(Nagpur and Amravati divisions)* - (Central Provinces and Old Berar Region)
- Marathwada Region - *(Aurangabad*
- Khandesh and Northern Maharashtra Region - *(Nashik Division)*
- Desh or Western Maharashtra Region - *(Pune division)* and
- Konkan Region - *(Konkan Division)* - (including, Mumbai City and Mumbai Suburban Area).

The state capital Mumbai City and Mumbai Suburban Area are the parts of the Konkan Division.

Border dispute

Maharashtra has a border dispute with neighbouring state of Karnataka over the district of Belgaum. Belgaum was incorporated into the newly formed Mysore state (now Karnataka) with the passage of the *States Reorganisation Act* (1956), which reorganised India's states along linguistic lines despite having about three-fourths of the total population.[55] speaking Marathi. Since then, Maharashtra has continued to claim the district. The case is currently awaiting a verdict in Supreme Court of India.

Principal urban agglomerations

Maharashtra has one of the highest level of urbanization of all Indian states.[56] The mountainous topography and soil are not as suitable for intensive agriculture as the plains of North India; therefore, the proportion of the urban population (42.4 per cent) contrasts starkly with the national averageveloping metro and many large towns.[57] Mumbai is the state capital with a population of approximately 15.2 million people. The other large cities are Pune, Nagpur, Nasik, Navi Mumbai, Thane, Amravati, Aurangabad, Kolhapur, Sangli and Solapur.

Mumbai, the Administrative Capital of Maharashtra, is also the largest city in India.

- **Mumbai:** Mumbai (including Thane and Navi Mumbai in its metropolitan area) is the financial and commercial capital of India and is the Administrative Capital of Maharashtra. It has the largest proportion of taxpayers in India and its share markets transact almost 70 per cent of the country's stocks. It offers a lifestyle that is rich, cosmopolitan and diverse, with a variety of food, entertainment and nightlife. Chhatrapati Shivaji International Airport (CSIA) in Mumbai is the biggest and the second busiest airport in India. The new airport, Navi Mumbai International Airport coming up at Panvel, Navi Mumbai will be the all modern and state of art facility airport in India. The city is India's link to the world of telecommunications and the Internet. VSNL (Now Tata Communications Limited) is the terminal point in India for all telephone and internet traffic. Mumbai is India's flagship port destination. It is also home to the Indian Navy's Western Command (INS).

- **Pune:** Pune, the second largest city in Maharashtra and the eighth largest in India, is the state's cultural and heritage capital with a population of 4.5 million people. About 170 km from Mumbai by road, Pune was the bastion of the Maratha empire. Under the reign of the Peshwas, Pune blossomed into a centre of art and learning. Several far reaching revenue and judicial reforms were also initiated in the city. Shaniwarwada, Saras baug, Aga Khan Palace, Parvati Temple, Khadakwasla Dam, Sinhgad are the most visited places by tourists in Pune. 'Ganeshotsav', a festival of Lord Ganesh is celebrated in Pune with lot of enthusiasm and worship. Pune is connected to Mumbai by the Mumbai-Pune Expressway. Pune also has very important military cantonments as well as the HQ of the Southern Command, the National Defence Academy, the Armed Forces Medical College, Pune, CME, and the Indian Air Force base at Lohegaon too. Pune is a major Information Technology hub of India as well as a foremost destination for Automobile manufacturing and component industry City.

Pune is the second largest city in the state.

"Zero Mile Stone" located at Nagpur.

- **Nagpur:** It is the third largest city in Maharashtra. The erstwhile capital of the Nagpur Province since 1853 in British India , which was in 1861 made the capital of Central Province , then in 1903, it was made capital of CP & Berar , then in 1935 it was made capital of a provincial assembly, with same name Central Provinces & Berar providing for an election by, Government of India Act , passed by British parliament. After independence, the "CP & Berar" was kept a separate entity with Nagpur as capital. In 1950 Nagpur became the capital of Madhya Pradesh. Nagpur was recommended capital of Vidarbha state by Hon. Fazal Ali commission for reorganisation of states. Nagpur was described by the first prime minister of India Mr. Jawaharlal Nehru, as " Heart of India". Nagpur is the nerve centre of Vidarbha (eastern Maharashtra), Nagpur - the *Orange City* as it is known - is located in the centre of the country and is also a geographical center of India, with a population of about 2.4 million people (2.1 Million as per census 2001). It is also Second Administrative Capital of Maharashtra. Nagpur is a growing industrial centre and the home of several industries, ranging from food products and chemicals to electrical and transports equipment. An international cargo airport, MIHAN is coming up in the outskirts of the Nagpur city. The Maintenance Command of Indian Air Force is located in Nagpur. The "Zero Mile Stone" or the geographical center of India is located in Nagpur. Deekshabhoomi, Sitabardi Fort, Ambazari Lake, Seminary Hills, Futala Lake, Dragon Palace Temple, Ramtek Temple, Khindsi Lake, Pench National Park are some of the tourist attractions in and around Nagpur.

Nashik is known for its pleasant climate co-existing with fast development.

- **Nashik:** It is one of the largest and highly industilised cities in the Maharashtra. One of the holy cities of the Hindu tradition, Nashik lies on the banks of the sacred river Godavari and has a population of about 1.6 million people. It is believed that Lord Rama, hero of the great Indian epic, the Ramayana, spent a major part of his exile here. Nashik is also a temple town, with over 200 temples. Nashik today is rapidly developing in ITs, industries, Pharmaceuticals and westernisation. It is also famous for its pleasant and cool climate. Nashik is also an educational hub.[58]
- **Aurangabad:** The city means "built by the throne", named after Mughal Emperor Aurangzeb), is a city in Aurangabad district, Maharashtra, India. The city is a tourist hub, surrounded with many historical monuments, including the Ajanta Caves and Ellora Caves, which are UNESCO World Heritage Sites, as well as Bibi Ka Maqbara. The administrative headquarters of the Aurangabad Division or Marathwada region, Aurangabad is said to be a 'City of Gates', as one can not miss the strong presence of these as one drives through the city. Aurangabad is also one of the fastest growing cities in the world.[59]

- **Sangli:** The 'Turmeric city of India' is the largest trade centre for turmeric in the country. Situated on the banks of river Krishna, Sangli-Miraj twin cities form the largest urban agglomeration in South Maharashtra. Sangli is famous for its grapes and Wine industry. Miraj is known for Indian classical musical instruments exported all over the world. Ganapati Temple of Sangli and its Ganesh Festival attract tourists from all over India. Sangli houses some renowned engineering and medical institutions. Density of hospitals in Sangli-Miraj twin cities is the highest in India. Sangli is now coming up as a major wind power generation hub. Sangli is well known for its sugar factories and dairy farms. Recently, large textile units are coming up around Sangli which is also witnessing developments in IT/ITES sector.

Aurangabad is a popular tourist destination.

Transport

Maharashtra has largest road network in India 267452[60] kilometers. National Highways in Maharashtra is 3688 kilometers [61] The Indian Railways covers most of the Maharashtra and is the preferred mode of transport over long distances. Almost the entire state comes under the Central Railways branch which is headquartered in Mumbai. Most of the coast south of Mumbai comes under the Konkan Railway. The Maharashtra State Road Transport Corporation (MSRTC) runs buses, popularly called ST for State Transport, linking most of the towns and villages in and around the state with a large network of operation. These buses, run by the state government are the preferred mode of transport for much of the populace. In addition to the government run buses, private run luxury buses are also a popular mode of transport between major towns.

Chhatrapati Shivaji International Airport is South Asia's largest aviation hub.

Mumbai has the biggest international airport in India with another coming up at Navi Mumbai. Pune has a civilian enclave international airport with flights to Dubai and Singapore, with plans on for a brand new greenfield International Airport. Aurangabad airport has recently been upgraded to an international airport with flights connecting to Jeddah. Other large cities such as Nagpur and Nashik are served by domestic airlines. Nashik has many flights To Mumbai and Soon Other Metros will be added after the construction of a new Airport at Nashik. Ferry services also operate near Mumbai, linking the city to neighbouring coastal towns. Other modes of public transport, such as a seven-seater tempo have gained popularity in semi-urban areas. Maharashtra has a large highway network. The Mumbai-Pune Expressway, the first access controlled tolled road project in India also exists within the state. Maharashtra has three major ports at Mumbai (operated by the Mumbai

The Chhatrapati Shivaji Terminus, is a key railway station and a UNESCO World Heritage Site.

Port Trust), the JNPT lying across the Mumbai harbour in Nhava Sheva, and in Ratnagiri, which handles the export of ores mined in the Maharastra hinterland.

Culture

Marathi is the official language of Maharashtra. Maharashtrians take great pride in their language and history, particularly the Maratha Empire, its founder Shivaji is considered a folk hero across India. About 80% of Maharashtrians are Hindu, and there are significant Muslim, Christian and Buddhist minorities. There are many temples in Maharashtra some of them being hundreds of years old. These temples are constructed in a fusion of architectural styles borrowed from North and South India. The temples also blend themes from Hindu, Buddhist and Jain cultures. A *National Geographic*[62] edition reads, "The flow between faiths was such that for hundreds of years, almost all Buddhist temples, including the ones at Ajanta, were built under the rule and patronage of Hindu kings." The temple of Vitthal at Pandharpur is the most important temple for the Varkari sect. Other important religious places are the Ashtavinayaka temples of Lord Ganesha, Bhimashankar which is one of the Jyotirling (12 important Shiva temples). Ajanta and Ellora caves near Aurangabad as well as Elephanta Caves near Mumbai are UNESCO World Heritage Sites and famous tourist attractions. Mughal architecture can be seen is the tomb of the wife of Aurangzeb called Bibi Ka Maqbara located at Aurangabad.

Kailash Temple in Ellora Caves.

Ajanta Caves

In 1708, the year following the death of Aurangzeb, Guru Gobind Singh the tenth spiritual leader of the Sikhs came over to Nanded, his permanent abode. He proclaimed himself the last living Guru and established the Guru Granth Sahib as the eternal Guru of the Sikhs. This elevates the reverence of Granth to that of a living Guru. A monument has been constructed at place where he breathed his last. Maharaja Ranjit Singh's endowment saw the construction of a beautiful Gurudwara at Nanded around 1835 AD. The Gurudwara features an imposing golden dome with intricate carvings and a breathtakingly beautiful artwork. It is known as Shri Huzur Abchalnagar Sachkhand Gurudwara.

Elephanta Caves

Maharashtra has a large number of hill, land and sea forts. Forts have played an important role in the history of Maharashtra since the time of the Peshwas. Some of the important forts in Maharashtra are Raigad, Vijaydurg, Pratapgad, Sinhagad. Majority of the forts in Maharashtra are found along the coastal region of Konkan.

Bollywood is based in Mumbai

Maharashtra, like other states of India, has its own folk music. The folk music viz. Gondhal Lavani, Bharud and Powada are popular especially in rural areas, while the common forms of music from the Hindi and Marathi film industry are favoured in urban areas.

The earliest instances of Marathi literature is by Sant Jnyaneshwar with his *Bhawarthadeepika* (popularly known as *Jnyaneshwari*). The compositions written during this period are spiritually inclined. The other compositions are by Sant Tukaram, Sant Namdev, and Sant Gora Kumbhar. The compositions are mostly in poetic form, which are called bhajans. These bhajans by saints are popular and part of day to day life. The modern Marathi literature has been enriched by famous poets and authors like P. L. Deshpande, Kusumagraj, Prahlad Keshav Atre and Vyankatesh Madgulkar. This literature has been passed on to the next generations through the medium of large numbers of books that are published every year in Marathi.

The world famous film industry Bollywood is in Maharashtra, located in the economic capital of India, Mumbai. The Marathi film industry was once placed in Kolhapur but now is spread out through Mumbai too. The pioneer of Indian movie industry, Bharat Ratna Shri Dadasaheb Phalke, producer & director V. Shantaram, B.R. Chopra, Shakti Samanta, Raj Kapoor, form a few names of the Hindi film fraternity, while writer, director, and actor P. L. Deshpande, actor Ashok Saraf, actor Laxmikant Berde, actor & producer, Sachin Pilgaonkar, Mahesh Kothare belong to the Marathi film industry. The early period of Marathi theatre was dominated by playwrights like Kolhatkar, Khadilkar, Deval, Gadkari and Kirloskar who enriched the Marathi theatre for about half a century with excellent musical plays known as Sangeet Naatak. The genre of music used in such plays is known as Natyasangeet. It is during this era of the Marathi theatre that great singer-actors like Bal Gandharva, Keshavrao Bhosle, Bhaurao Kolhatkar and Deenanath Mangeshkar thrived.

Some of the popular Marathi television channels are Star Majha, Zee Talkies, Zee Chovis Taas, Mi Marathi, DD Sahyadri, Zee Marathi, ETV Marathi, and Saam Marathi which host shows ranging from soap operas, cooking and travel to political satire and game shows.

The cuisine of Maharashtra varies according to the region of Maharashtra. The people of the Konkan region have a chiefly rice based diet with fish being a major component, due the close proximity to the sea. In eastern Maharashtra, the diet is based more on wheat, Jowar and Bajra. Puran Poli, Bakarwadi, plain simple Varan Bhat (a dish cooked with plain rice and curry), and Modak are a few dishes to name. Chicken and mutton are also widely used for a variety of cuisines. Kolhapuri Mutton is a dish famous for its peculiar spicy nature.

DY Patil Stadium in Navi Mumbai

Women traditionally wear a nine yard or five yard sari and men a dhoti or pajama with a shirt. This, however, is changing with women in urban Maharashtra wearing Punjabi dresses, consisting of a Salwar and a Kurta while men wear trousers and a shirt.

The cricket craze can be seen throughout Maharashtra, as it is the most widely followed and played sport. Kabaddi and hockey are also played with fervor. Children's games include Viti-Dandu (Gilli-danda in Hindi) and Pakada-pakadi (tag).

Hindus in Maharashtra follow the Shalivahana Saka era calendar. Gudi Padwa, Diwali, Rangapanchami, Gokulashtami and Ganeshotsav are some of the festivals that are celebrated in Maharashtra. Ganeshotsav is one of the biggest festival of Maharashtra which is celebrated with much reverence and festivity throughout the state and has since some time become popular all over the country. The festival which continues over ten days is in honour of Ganesha, the god of learning and knowledge. A large number of people walk hundreds of kilometers to Pandharpur for the annual pilgrimage in the month of Ashadh.

See also

- Districts of Maharashtra
- Geography of Maharashtra
- History of Maharashtra
- Marathi language
- Marathi people
- Tourism in Maharashtra

External links

- Maharashtra State Blood Transfusion Council [63]
- Government of Maharashtra [64]
- Public Works Department Maharashtra State Official Website [65]
- Maharashtra travel guide from Wikitravel

References

[1] "Maharashtra Tourism: Trivia" (http://www.maharashtratourism.gov.in/MTDC/HTML/MaharashtraTourism/Trivia.html). *Official website of Maharashtra Tourism*. Government of Maharashtra. . Retrieved 2007-07-16.

[2] Palkar, A.B (2007), *Report of One Man Commission Justice A.B.Palkar: Shri Bhaurao Dagadu Paralkar & Others V/s State of Maharashtra* (http://www.maharashtra.gov.in/pdf/VOLUME-I.pdf), **I**, p. 41, , retrieved 2007-07-16

[3] ""Maharashtra", Government of India, Ministry of Home Affairs, National Informatics Centre. (NIC)" (http://mha.nic.in/States\maharashtra.pdf) (PDF). . Retrieved 2007-05-01.

[4] http://maharashtra.gov.in

[5] http://en.wikipedia.org/wiki/List_of_Indian_states_by_GDP

[6] "Introduction to Maharashtra Government" (http://www.maharashtraweb.com/Government/intro.asp). Maharashtraweb.com. . Retrieved 2008-10-31.

[7] (http://www1.worldbank.org/wbiep/decentralization/saslib/urban reforms.ppt)

[8] "India - Maharashtra" (http://www.worldbank.org.in/WBSITE/EXTERNAL/COUNTRIES/SOUTHASIAEXT/INDIAEXTN/0,,contentMDK:20951183~pagePK:141137~piPK:141127~theSitePK:295584,00.html). Worldbank.org.in. . Retrieved 2008-10-31.

[9] GDP of Indian states (http://mospi.nic.in/6_gsdp_cur_9394ser.htm)

[10] ^ >^ Thoughts on linguistic states. http://www.ambedkar.org/ambcd/05C.%20Thoughts%20on%20Linguistic%20States%20PART%20III.htm

[11] The Tribes and Castes of the Central Provinces of India - Volume IV of IV (http://www.gutenberg.org/files/20668/20668-h/20668-h.htm#d0e6680)

[12] Sir H. Risley's India Census Report (1901), Ethnographic Appendices, p. 93.

[13] An inscription at Naneghat describes Vedishri as a very brave king and the lord of Dakshinapatha (Deccan). Mirashi, *Studies in Indology*, vol. I, p. 76 f.]

[14] "MAHARASHTRA TOURISM, The Official Website of Maharashtra Tourism Development Corporation, Govt. of India" (http://www.maharashtratourism.gov.in/). Maharashtratourism.gov.in. . Retrieved 2008-10-31.

[15] http://www.iloveindia.com/indian-heroes/bal-gangadhar-tilak.html

[16] New Page 1 (http://www.wii.gov.in/envis/envis_pa_network/page_maharashtra.htm)

[17] Maharashtra economy soars to $85b by 2005 (http://specials.rediff.com/money/2009/mar/31slide13-indias-top-ten-debt-ridden-states.htm)

[18] Maharashtra debt climbs to 36 per cent of GDP (http://specials.rediff.com/money/2009/mar/31slide13-indias-top-ten-debt-ridden-states.htm)

[19] "Maharashtra achieves Rs 810 Cr revenue surplus- Finance-Economy-News-The Economic Times" (http://economictimes.indiatimes.com/News/Economy/Finance/Maharashtra_achieves_Rs_810_cr_revenue_surplus/rssarticleshow/2983478.cms). Economictimes.indiatimes.com. . Retrieved 2008-10-31.

[20] "Maharashtra bifurcation will fetch additional Rs30,000 crore annually" (http://www.dnaindia.com/mumbai/report_maharashtra-bifurcation-will-fetch-additional-rs30000-crore-annually_1338556). Dnaindia.com. . Retrieved 2010-01-28.

[21] http://www.pppinindia.com/business-opportunities-maharashtra.asp

[22] http://www.business-standard.com/india/news/mumbai-airport-plans-rs-2280-cr-investment-this-fiscal/94602/on

[23] "Identification of suitable sites for Jatropha plantation in Maharashtra using remote sensing and GIS" (http://www.unipune.ernet.in/dept/geography/vhdeosthali_files/jatropha.htm). University of Pune. . Retrieved 2006-11-15.

[24] "A model Indian village- Ralegaon Siddhi" (http://edugreen.teri.res.in/explore/renew/rallegan.htm). . Retrieved 2006-10-30.

[25] "Maharashtra Airport Development Company Limited" (http://www.madcindia.org/aboutmadc.htm). *www.madcindia.org*. www.madcindia.org. . Retrieved 2008-05-14.

[26] "Maharashtra Airport Development Company Limited" (http://pib.nic.in/archieve/factsheet/2005/fscivil2005.pdf). *Press Information Bureau and Ministry of Civil Aviation*. pib.nic.in. . Retrieved 2008-01-29.

[27] ""Nagpur stakes claim to lead boomtown pack"" (http://www.indianexpress.com/story/3713.html). The Indian Express. . Retrieved 2006-06.

[28] "Mihan is biggest development" (http://timesofindia.indiatimes.com/Cities/Nagpur/Mihan_is_biggest_development_project/articleshow/2065727.cms). *timesofindia.indiatimes.com*. timesofindia.indiatimes.com. . Retrieved 2007-05-22.

[29] "Twelfth Finance Commission" (http://fincomindia.nic.in/). Finance Commission of India. . Retrieved 2006-09-19.

[30] http://www.educationinfoindia.com/maharashtradir.htm

[31] http://www.projectsmonitor.com/detailnews.asp?newsid=2076

[32] "History" (http://www.mu.ac.in/History.html). University of Mumbai. . Retrieved 2009-06-09.

[33] "IIT flights return home" (http://www.dnaindia.com/report.asp?NewsID=1070723). *Daily News and Analysis (DNA)*. 2006-12-22. . Retrieved 2009-06-09.

[34] "About the Institute" (http://www.vjti.ac.in/home_about.asp). Veermata Jijabai Technological Institute (VJTI). . Retrieved 2009-06-09.

[35] "Admission process for autonomous engg colleges to start today" (http://www.expressindia.com/latest-news/admission-process-for-autonomous-engg-colleges-to-start-today/321286/). Indian Express Group. 2008-06-11. . Retrieved 2009-06-09.

[36] "About University" (http://sndt.digitaluniversity.ac/Content.aspx?ID=7&ParentMenuID=7). SNDT Women's University. . Retrieved 2009-06-09.

[37] Bansal, Rashmi (2004-11-08). "Is the 'IIM' brand invincible?" (http://in.rediff.com/getahead/2004/nov/08rash.htm). *Rediff News* (Rediff). . Retrieved 2009-06-09.

[38] "Sydenham College: Our Profile" (http://www.sydenham.edu/our_profile.html). Sydenham College. . Retrieved 2009-04-26.

[39] "About The Government Law College" (http://www.glc.edu/incept.asp). Government Law College. . Retrieved 2009-04-26.

[40] Martyris, Nina (2002-10-06). "JJ School seeks help from new friends" (http://timesofindia.indiatimes.com/articleshow/24305727.cms). *The Times of India*. . Retrieved 2009-05-13.

[41] "University ties up with renowned institutes" (http://www.dnaindia.com/report.asp?NewsID=1065998). *Daily News and Analysis (DNA)*. 2006-11-24. . Retrieved 2009-06-09.

[42] "CIRUS reactor" (http://www.barc.ernet.in/webpages/reactors/cirus.html). Bhabha Atomic Research Centre (BARC). . Retrieved 2009-05-12.

[43] http://www.indiaeducation.ernet.in/insitutions/PROFILENEW.ASP?no=U00565

[44] "Mahapopulation" (http://www.maharashtra.gov.in/english/ecoSurvey/ataglanc.pdf) (in Marathi) (PDF). *Census of India*. www.maharashtra.gov.in. . Retrieved 2008-06-04.

[45] http://www.maharashtra.gov.in/english/ecoSurvey/ecoSurvey2005-06/eng/cha_3e.p

[46] http://festivals.tajonline.com/ganesh-chaturthi.php

[47] http://www.krishnajanmashtami.com/ceremony-dahi-handi.html

[48] दसरा FAQ - All about Dasara in Maharashtra (http://sanatanhindudharma.blogspot.com/2008/10/blog-post_06.html)

[49] Dasara Rituals in Maharashtra (http://www.hindupad.com/2009/09/simollanghan-â-dasara-ritual-in-maharashtra/)

[50] Shepherd, P. 111 *Gurus Rediscovered:*

[51] . http://us.asiancorrespondent.com/Indianomics/2009/01/10/shirdi-sai-baba-beats-credit-crunch-to-become-the-2nd-richest-temple-in-the-country.. http://news.zakhas.com/2009/01/shirdi-sai-baba-temple-second-richest-in-country/.

[52] "2001 Census" (http://www.censusindia.gov.in/Census_Data_2001/Census_Data_Online/Language/Statement3.htm). Ministry of Home Affairs, GOI. . Retrieved 2008-10-31.

[53] List of districts and divisions (http://www.maharashtra.gov.in/english/mahInfo/)

[54] Maharashtra - Facts and Figures (http://www.maharashtra.gov.in/english/mahInfo/state.php)

[55] Jaishankar Jayaramiah (21 November 2005). "Karnataka caught in 'language' web" (http://www.financialexpress.com/fe_full_story.php?content_id=109230). The Financial express. . Retrieved 2006-11-01.

[56] Level of Urbanisation (http://www.urbanindia.nic.in/moud/urbanscene/levelofurbanisation/main.htm)

[57] http://www.maharashtraweb.com/majorcities.asp

[58] http://www.citypopulation.de/India-Maharashtra.html#Stadt_agglo All towns and agglomerations in Maharashtra of more than 20,000 inhabitants.

[59] 11 Indian cities among worlds fastest growing. (http://timesofindia.indiatimes.com/news/india/11-Indian-cities-among-worlds-fastest-growing/articleshow/2481744.cms)

[60] http://morth.nic.in/writereaddata/sublinkimages/251.html

[61] http://www.mahapwd.com/statistics/default.html

[62] January 2008, VOL. 213, #1
[63] http://mahasbtc.aarogya.com/index.php
[64] http://www.maharashtra.gov.in/index.php
[65] http://www.mahapwd.com/

Article Sources and Contributors

Bombay Explosion (1944) *Source*: http://en.wikipedia.org/w/index.php?oldid=333589235 *Contributors*: Apyule, Ardfern, Ashish Sampat, Bryan Derksen, Carcharoth, Cdang, DabMachine, Fredrik, Gaius Cornelius, Guptadeepak, Hedgehog, Hmains, Hugo999, Icairns, Jaraalbe, Kpjas, Lightmouse, Murderouspigeon, Mw66, Nichalp, Ning-ning, Nishkid64, PFHLai, Pearle, Piano non troppo, Rama's Arrow, Sasateam, Seano1, Skookum1, SlaveToTheWage, Suyogaerospace, TransporterMan, Wayward, Whpq, Wik, Yash Momaya, Yash512, [212], 10 anonymous edits

Victoria Dock *Source*: http://en.wikipedia.org/w/index.php?oldid=354347421 *Contributors*: Andypasto, Chubbles, Jaraalbe, Lenthe, Peter Ellis, Shantavira, Shortfatlad, Snowy 1973, The Anome, Zaian, 4 anonymous edits

Mumbai *Source*: http://en.wikipedia.org/w/index.php?oldid=363869109 *Contributors*: 190319m9, 1excalibur, 200.191.188.xxx, 203.109.250.xxx, 205gti306gti, 21655, A i s h2000, ABF, AI009, ASOTMKX, Aaftabj, Aakashshah123, Aakkshay, Aam422, Aaroncrick, Abecedare, Abhayrk, Abhijit borude, Abhijitsathe, Abhishek 619, Abhishekjparmar, Acalamari, Ackulkarni, Actuszeus, Adam.J.W.C., Adhishb, Advaitrocking, AgarwalSumeet, Agceltics, Ahmed27, Ahoerstemeier, Ahuskay, Aiyer, Ajay259, Ajayr22, Ajayraote, AjitDongre, AjitPD, Akar2, Aksi great, AlainaUCSD, Alamandrax, Alanraywiki, Alansohn, Aleenf1, AlexNebraska, Alexius08, Algesh, Alren, Alsandro, Amartyabag, Ambarish, Ambuj.Saxena, Amirki, Amitdotchauhan, Amol.Gaitonde, Amsterdam360, Anchitk, AndonicO, Andre Engels, AndrewHowse, Andrewlp1991, Andrewman327, Angela, AnimalExtender, Anirbandas, Ankitbhatt, Anonymous12321, Anshuk, Antaeusrentacar, Antandrus, Antares784, Anthony Appleyard, Anupam, Anuraagvaidya, ApolloCreed, Appraiser, Arch dude, Archanajnu, AreJay, Arfatazmi, Arjun024, Arnoldsty, Aronjr41, AroundTheGlobe, Art LaPella, Artaxerxes07, Artaxiad, Arun11, Asahai.248, Ascend, Asenine, Ashurockstarboy, Ashwatham, Athaenara, Augur, Avala, Awnabbas, AxelBoldt, Azimuth1, BBird, BD2412, Baa, Backslash Forwardslash, Bala 207, Balthazarduju, Balwinderdeep, Banes, Banzaicat, Barfooz, Bariajatin, Bart133, Batram, Bazonka, Bcorr, Beetstra, Behzadaltaf, Beland, Belasd, Bentley4, Bergsten, BernardM, Bernardlabs, BethelRunner, Bharan cse, Bhaskarswaroop, Big Adamsky, Bill37212, Billposer, Biruitorul, Bkell, Black Kite, Blackjays, Blake-, Bloodshedder, Bluedenim, Bluezy, Bnitin, Bobblewik, Bobet, Bobo192, Bobsheth, Bodybagger, Bogdan, Bogomolov.PL, Bollywoodbombay, Bomlove, Bongwarrior, Boothy443, BorgQueen, Boston, BostonPunekar, Boyfriendtogirlsaloud, Brandmeister, Bred, Brhaspati, Brighterorange, Brusegadi, Bryan Derksen, Bsadowski1, By78, CPMcE, CRKingston, Caknuck, CalJW, Camw, Can't sleep, clown will eat me, Canaima, Cantus, CapitalR, Capricorn42, CarTick, Cargoking, Carsos1992, Cdc, Cenarium, Cgs, Chameleon, Chanakyathegreat, Chance Harper, Chengiz, Cheraz, Chewwy225, Chimanrao, Chinmay26r, ChiragPatnaik, Chirags, Chris Roy, CieloEstrellado, Cirt, Clark89, Clawed, Closedmouth, Cmdrjameson, Colibri37, CommonsDelinker, Confuzion, Conversion script, Coolazhar, Coollemonade, Cosmic Latte, Cpbaherwani, Crambane, CrashCart9, Crazysoul, Crazyvas, Criticalthinker, Cromwellt, Crusoe8181, Crystalmaze 230583, Cs-wolves, Cst17, Cukorka22, Curps, Cyberdiablo, Cyclopaedia, D6, DBGrover, DJ Clayworth, DMG413, DO'Neil, DaGizza, Dabomb87, Dalekrabe, Daniel, Danny-w, Dantadd, DaronDierkes, Darrendeng, David R. Ingham, David matthews, Davidcannon, Davshul, Dbachmann, Deavenger, Debresser, Deepak, Deepak D'Souza, Delirium, Delldot, Deltabeignet, DerHexer, DerechoReguerraz, Derek.cashman, Desiphral, Devanjedi, Deville, Dewan357, Dhruv9876543210, Diberri, Dide14, Dilcsi, Dilli Billi, Discospinster, Dksidana, Dmrcjess, DocWatson42, Doczilla, Donarreiskoffer, Dontrustme, Doorvery far, Dorelal, Dr. Blofeld, Drmies, Dryman, Dsouzamarshall, Dubaiproperty, Dudecon, Duskrider, Dwaipayanc, ESkog, Easydriveforum, Ed Poor, Edcolins, Effer, Eggman64, Ekabhishek, Ekotkie, El C, Eladush, Elassint, Electronz, Elinnea, Elockid, Enigma Blues, Enihar, Enthusiast10, Erode, Essell sons, Estel, Eubulides, Everyking, Everything counts, Evil Monkey, Ewlyahoocom, Extra999, Faisalzar 29, Fan-1967, Farkas János, FayssalF, Fedayee, FindAWayOrMakeOne, Finetooth, Fjmustak, Fooferoni101, Foulis, Fourfirefoxes, Fowler&fowler, Fox, Frau Holle, Freakofnurture, Fredrik, FreplySpang, Fruge, Fudoreaper, Fundamental metric tensor, Fusion07, Futurebird, Fuzheado, Fæ, GDibyendu, GNRY09, Gabbhh, Gadget850, Gaius Cornelius, Gaius Octavius Princeps, Gajamukhu, Ganeshk, Garytown, Gaurav, Gautamj, Gdarin, Geeteshgadkari, Gelo71, Gene Nygaard, Generalboss3, Geni, GeorgeGill100, GetLinkPrimitiveParams, Ghushe, Gimmetrow, Girgirya, GlassCobra, Gmaxwell, Gogo Dodo, Gold heart, Gomathishankar1975, Gopalvsh, Gppande, GraemeL, Green Giant, GregorB, Grenavitar, Gsklee, Guesme, Gurudev123, Gyan, Hadal, Hajor, Hamako, Hamiltonstone, Hammersoft, Hamtechperson, Handbook3, Handlins, Happylobster, Hardik jadeja, Hardouin, Hari, Harsha363, Harshadkhandare, Harshits, Helloanand, Hellooooo, Heman, Hemanshu, Hemendra, Henry Flower, Hfastedge, Hintha, Hinto, Hkelkar, Hmains, Hnsampat, Holy Ganga, Hometech, Hongooi, Hu12, Huniebunie, Huseyx2, Hux, Hwbonkers, IAF, Iamvolunteer, Icairns, Ifmanish, Imhunt, ImpuMozhi, India Rising, Indianboy1, Indianhilbilly, Indon, Indoresearch, Indscribe, Information-Line, Infrogmation, Interlingua, Ioeth, Ipigott, Iridescent, Irutavias, Ishan, Itz akshay, Itzsundar, J.delanoy, J04n, JPD, JaGa, Jackol, Jacksav, Jacoplane, Jaganath, Jagged 85, Jainrajat11, Jainuday, Jalimelys88, Jalovaalea, Jan van Male, Jana337, Jane McCann, Janx Spirit, Jarry1250, Jasepl, Java13690, Jaxl, Jay, Jaymerchi, Jeandré du Toit, Jeff G., JeffreyN, Jellogirl, Jeroen, Jeronimo, Jerryseinfeld, Jguk, Jguk 2, Jim, Jkeene, Jlin, Joaopais, Joe3600, Joedjemal, John, Johnhardcastl, Johnian144, Johnvasai, Jojit fb, Jonathan.s.kt, Jondr12, Joowwww, Jose Ramos, Joseph Solis in Australia, Jovianeye, Joyson Konkani, Julius Nayak, Justinitalia, Justmihir, Juzer, KCinDC, KJS77, KMANOJ06, KNM, KRS, Kaal, Kafka Liz, Kairos, Kamalkantchourey, Kamielschwartz, Kandarp.desai82, Kanduparee, Karenjc, Karish, Karmosin, Karthik.coep, Karyasuman, Kash1234, Kashifkhusro, Kate, Katimawan2005, Kbdank71, Kbh3rd, Kechi, KeithCHoffman, Kelisi, Kelly, Kelvinng90, Kensplanet, Kentem, Kesangh, Kevin, Khcf6971, Khivi, Khoenr, Khoikhoi, King Zebu, King of Hearts, King747, Kingboyk, Kingfisherisgay, Kinu, Kipala, Kittoo, Kkm010, KnowledgeHegemony, KnowledgeHegemonyPart2, KnowledgeOfSelf, Kokarako Gumango, Kola1991, Kparmar24, KrakatoaKatie, Kristenq, Krithix, Kshitij85, Kukini, Kulasman, Kunalkariwala, Kungfuadam, KuwarOnline, Kwamikagami, Kwsn, L.vivian.richard, LA2, LachlanA, Lagalag, Lalit Jagannath, Lambiam, Lankiveil, Larry V, Last Emperor, Latitude0116, Le Canari, Leafyplant, Leandrod, Lear 21, Lear's Fool, LeaveSleaves, Lectonar, Leopart, Light48, LightAnkh, Lightmouse, LilHelpa, LittleOldMe old, Livajo, Llort, Local hero, Loganberry, Loginfirstname, LonelyMarble, Longhair, Lord Emsworth, Lord Voldemort, LordSimonofShropshire, Loren.wilton, Lostintherush, LouisSS13, Lova Falk, Lucio Mas, Luk, Lyght, Lzur, M m hawk, M3taphysical, MBorghese, MER-C, MJCdetroit, Maddie!, Maddyr, Madscientistjaidev, Magicalsaumy, Mah exp, Maha Mantri, Maharashtraexpress, Mahawiki, Mailmustu, Major Bonkers, Makks2010, Malayali youth, Malleus Fatuorum, Mamboraj, ManasGupta85, ManasShaikh, Manaspunhani, Mangald, Manish367, Manuel Anastácio, Marathipremi, Marek69, Mark0687, Markus Schmaus, Martandsinhparmar, Martin.Budden, Martin451, Marwek, Massimillio, Matdrodes, Matilda, Matt.T, Matthewhicks93, Mattisse, Mauls, Mav, Maxim, MaximvsDecimvs, Maxis ftw, Maya, Mayank.hc, Mayankbid, Mayankgates, Mayur.thakare, Mayuresh.kathe, Me-la-pelan, Meb53, Megaman24, Meghanand, Mendaliv, Mercunis, Metarhyme, Mh12, Michael Devore, Michael G. Davis, Michael Snow, MichaelTinkler, MichiganCharms, Micke-sv, Mike Rosoft, Mikey79au, Millosh, Mimic2, Minesweeper, Mjolnir1984, Mkweise, Moe Epsilon, Mohanpn, Mohitkuk, Moncrief, Mono, Monster eagle, Montrealais, Mouleesha, Mowgli, MrChile, Mrdrift, Mrehere, Mrtag, Mrtpolice, Mspatnaik, Muchomank, Mudkid, Mumbai terrorist bomber, Mumbaicha, Mumbaikid, Murali83, Murdernacho, Muriel Gottrop, Murtasa, Mushroom, Musicpvm, Mustaqbal, Mutiqb, Myanw, Myheartinchile, Mzsatish, NCrane, NE2, Nachoman-au, Nakon, Naveenbm, NawlinWiki, Nburden, Ndenison, Neilf11, Neko-chan, Neoform, Neptomann, Net2008, Netzarp, Neutrality, Neverquick, Neviselevate, New Rock Star, NewEnglandYankee, Nhajaj, Nichalp, Nicke L, Nicsa, Nikhil Govalkar, Nikkul, Ninadhardikar, Ninadism, Nirav.maurya, Nirbhaykanoria, Nirjosho, Nirvana888, Nishant12, Nishkid64, Nitajk, Niteowlneils, Nitincoolchamp, Nitinjamdar, Nitnaga, Njaelkies Lea, Nk, Nku, Nmpenguin, Npcrazy98, Npindia, Nskamat, Nthakk01, Nubiatech, NuclearWarfare, NupeWD1993, Nuttycoconut, Ny512, O1ive, Observ, Ohnoitsjamie, Oldscientist, Oliver Pereira, OneHappyHusky, Op. Deo, Optakeover, OrangeINDIA, Ortolan88, Ottava Rima, Ottre, OverlordQ, OwenX, OzWhiz, P.K.Niyogi, PDH, PL290, PS2pcGAMER, PZFUN, Paddu, Page Up, Paine Ellsworth, Palazzi 95, Papertree, Paragchitre, Patrick, Paul-L, Paulinho28, Paxsimius, Peachey88, Pearle, Pediaindia, Pepsidrinka, Per Honor et Gloria, Personnne, Peter Horn, Peter I. Vardy, Petronas, Pharaoh of the Wizards, Phatmastafunk, PhilKnight, Philip Trueman, Philycwalk, Pigsonthewing, Pinkadelica, Pizzadeliveryboy, Planemad, Plastictv, Plasticup, Plastikspork, Playethic, Pmsyyz, Poccil, Polaron, Portress, Possum, Postglock, Prachit, Pradeepbv, Pradeshhimachal, Pranab.salian, PranksterTurtle, Prasadksap, Prasanaik, Pratheepps, Priyanath, Prodego, Proofreader77, Proteus, Psresh, Psurajit, Ptcamn, Pumelo, Pungimaster, Punitpankaj, Pwarrior, Pyromaniac131313, Quadell, Quebec99, R saji kumar, RDSweet, RG104, RG2, RRSD, Radiant!, RafaAzevedo, Rafaelgr, Ragib, Raheelansari, Rahul meg, Rahulgo4love, Rait, Rajasekaran Deepak, Rajeshroshan, Rajithmohan, Rakov, Rama's Arrow, Ramrao, RamunasM, Ran, Randomblue, Ranga27, Rantingraj, Ratzd'mishukribo, Rav.pat.2007, Raven in Orbit, Ravikiran r, Ravn, Raymond78, RayquazaDialgaWeird2210, Razorflame, Razzing, Rebecca, Rebelduder69, RedWolf, Redfarmer, Redrose64, Redsox24123, Redtigerxyz, RegentsPark, Reliancepowercoin, Renaissancee, Restname, RetiredUser2, Rettetast, Rgoogin, Rhcp fanin, Rhobite, Riana, Rich Farmbrough, Richard Woods, RickK, Ritchy, Rjwilmsi, Rl, Rmathur, Robby, RobertG, Robertson-Glasgow, Robin klein, Rocket000, Rohan.kurani, Rohirok, Rohit Vaidya, Rohitbd, Rohittikmany, Romanwarrior, Rosh2610, Roshan baladhanvi, Rq88187, Rrjanbiah, Rror, Rsrikanth05, Rukateja, Rumpelstiltskin223, RupertMillard, RussBlau, Russavia, Rvbhatia, Ryan Postlethwaite, RyanEberhart, Ryanduck35, S h i v a (Visnu), S3000, SBC-YPR, SI4, SMC, SNIyer12, SPUI, ST47, SUL, Sachindole, Sag777, Saga City, Saheballah, Sakshat, Sam Hocevar, Sam Korn, Sammysam, Samsoumi95, Samul, Sanfranman59, Sanfy, Sanilkeni, Sanjay ach, SanjayTilaiyan, Sanket ar, Sannse, Sansonic, Sanya, Sapan gajjar, Saravask, Sardanaphalus, Saros136, Sarvagnya, Sashidharg, Sasikumar10, Saturn star, Sayalithebest, Sceptre, Sct72, Seivad, Send2parvez, Sevenland, Sfahey, Shadowjams, Shanes, Shantanuo, Shantanutandon, Shashidc, Shizhao, Shj95, Shoeofdeath, Shorerydr10, Shouriki, Shyam, Siddhant, Siddharth Prabhu, Sidonuke, Signalhead, Signatius, Sikandarji, Silkyroad, SimonArlott, SimonP, Simplyj, Sir Nicholas de Mimsy-Porpington, Siroxo, Skcpublic, SlaveToTheWage, SmartGuy, Smartdashboard, Smartdust, Smartinfoteck3, Snickster 02, Soilguy5, Solar20, Soman, Soulpatch, Soumyadipc, SpacemanSpiff, Speedevil, Spideraptor, Splurge, Sprashu, SqueakBox, Squiquifox, Srikeit, Srini81, Srinikasturi, Srinivara, Ssharma55, Stateofart, Stegop, Stephen, Stephen Turner, SteveRwanda, Stupid Corn, Subcontinent, Suhanasoft, Sukh, Sumedh.hoskote, Sundar, Sunil N, Supasheep, Superdude99, Suthra, Suyogaerospace, SuzanneKn, Sverdrup, Svishwanathan, Swaminworld, Switzpaw, SyedNaqvi90, Sylvain1972, TShilo12, TXiKi, Tabletop, Tal Celes, Tanooj, Tanthalas39, Tanvirg, Tariqabjotu, Tawker, Tehlulz54, Tejas81, Tempodivalse, Thaejas, The Evil Spartan, The Ogre, The Vivacious Vivienne VaVane, The free one, The undertow, The wub, TheCatalyst31, TheEgyptian, Thegroundsquirrel, Thehalfone, Thesmothete, Thewarriorrrr, Thingg, ThomasPhilip.jr, Thunderboltz, TigerShark, Timir2, Timsdad, Timwi, Tjmayerinsf, Tjrudebeck, Tkynerd, Tmopkisn, Tobby72, Tobias Conradi, Tom Radulovich, Tony Sidaway, TonyTheTiger, Toprohan, Tpbradbury, Trafford09, Trakesht, Trek011, Trovatore, Trusilver, TryCatchDenz, Tsrajesh, Tuncrypt, TutterMouse, Tvmpolice, Twistkhan, U.prasad, UW, Universe inside, Urhixidur, Urnonav, Usyerramilli, Utcursch, UtherSRG, V das, Vadakkan, Vancouverguy, Vasaijohn, Vatech09, Vdeolali, Vedant, Vegaswikian, Velho, VeryVerily, Veryhuman, Vfitzpatrick, Vhankate, Victoria Eleanor, Vidveng, Villysap101, Vinay Pednekar, Violetriga, Vivo78, Viyu5, Vkvora2001, Vr5106, Vssun, WISo, Walrabbit, Walter Görlitz, Ward3001, Warrennoronha, Warunshahc, Water Fish, Weissmann, Whbonney, WhisperToMe, Widders, Wik, Wikeditor22, Wikiality123, Wikidea, Wikideewana, Wikidemon, Wikipedia brown, Wild Pansy, Wimt, Wknight94, Wmahan, Woohookitty, Wooky70, World, World8115, Worldtraveller, Xavier449, Xchary, Xman, Xone4thekidsx93, YUL89YYZ, Yahel Guhan, Yamamoto Ichiro, Yamla, YellowMonkey, Yerpo, Yoddal, Yogb21, Yogesh Khandke, Yogeshtk, ZakuSage, Zangar, Zeno Gantner, ZiyadMadon, Zoe, Zoicon5, ZooFari, Zoomzoom, Zora, Zulfikkur, Zundark, Zzuuzz, ~shuri, రవిచందర్, 每日飞龙, 2589 anonymous edits

SS Fort Stikine *Source*: http://en.wikipedia.org/w/index.php?oldid=91017267 *Contributors*: Apyule, Ardfern, Ashish Sampat, Bryan Derksen, Carcharoth, Cdang, DabMachine, Fredrik, Gaius Cornelius, Guptadeepak, Hedgehog, Hmains, Hugo999, Icairns, Jaraalbe, Kpjas, Lightmouse, Murderouspigeon, Mw66, Nichalp, Ning-ning, Nishkid64, PFHLai, Pearle, Piano non troppo, Rama's Arrow, Sasateam, Seano1, Skookum1, SlaveToTheWage, Suyogaerospace, TransporterMan, Wayward, Whpq, Wik, Yash Momaya, Yash512, [212], 10 anonymous edits

Prince Rupert, British Columbia *Source*: http://en.wikipedia.org/w/index.php?oldid=363660536 *Contributors*: Alexbatko, Andres, Andy Christ, Angela, Antiuser, Arct, Bearcat, Beland, Black Tusk, Bobblewik, Cadillac, Can't sleep, clown will eat me, CaseyPR, Cfailde, Chairman S., Circeus, Cmr08, Crcann, D6, Dab235, David Kernow, Davidjames25, Dogbreathcanada, Earl Andrew, Eeee, FairFare, Flauto Dolce, G2bambino, Gadfium, Geo Swan, Geomr, GlassCobra, Grafen, Greg Salter, Ground Zero, Hamm1337, Indefatigable, Joelfurr, Johntwrl, Jonathan.s.kt, Keefer4, Kenn1987, Kurieeto, Kyle1278, Lajkonik, LegolasGreenleaf, Leslie Mateus, Lights, Luckyluke, M@sk, MJCdetroit, Mathpianist93, Matthuston, Mburns1655, MechBrowman, Miesianiacal, Mig29, Mindmatrix, Mlaffs, Olzver, PastorBobinAustin, Paul Erik, Peter Horn, Philippe, Plasma east, Polylerus, Quadell, Qyd, Radagast83, RedWolf, Rich Farmbrough, Rich257, Rjwilmsi, Robeby70@hotmail.com, Rwxrwxrwx, Sammalin, Scriberius, Sdrinkwater, Siubhan, Skookum1, Skunyote, Steam5, Struway, TL789, Tartessos75, Timc, Usgnus, Vagary, WadeSimMiser, Wakemp, Wavelength, Welsh, Wiki Roxor, Wknight94, Woohookitty, Zeus1234, 112 anonymous edits

Fort Stikine *Source*: http://en.wikipedia.org/w/index.php?oldid=363239232 *Contributors*: Backspace, Bazonka, CambridgeBayWeather, Davehi1, DuncanHill, EeepEeep, Kwiki, Mjroots, Pfly, Rich Farmbrough, Skookum1, 2 anonymous edits

Colaba Observatory *Source*: http://en.wikipedia.org/w/index.php?oldid=342711277 *Contributors*: Akarkera, Bazonka, DuncanHill, Ekabhishek, Michael Ronayne, PamD, PigFlu Oink, Salih, V1adis1av, WereSpielChequers

Hudson's Bay Company *Source*: http://en.wikipedia.org/w/index.php?oldid=363492081 *Contributors*: 0, 193.133.134.xxx, 6.7, 96.149, AEMoreira042281, Ab762, Aboluay, Aboutmovies, Adam Bishop, Akamad, Albrecht, Alex43223, AlexRampaul, Alexwcovington, Ali@gwc.org.uk, Allstarecho, Amii, Andrewpmk, Angelique, Anonymous editor, Arctic.gnome, ArkansasTraveler, Artene50, Arthana, Artyboy, Ary29, Avant Guard, Awesome60, BaddeckAcademy, Basho, Bcorr, Bearcat, Beast9989, Benw, Best O Fortuna, Bigrich, Bill37212, Bkonrad, Bobblewik, Bobo192, Boing! said Zebedee, Bradjamesbrown, Brian Crawford, Bryan Derksen, Bryansit83, Bssc81, Burgundavia, By Little Old Me, CJLL Wright, CWood, Cahk, Calliopejen, CambridgeBayWeather, Canmoore, Canterbury Tail, Caster23, Cdills, Centrx, Cgweiss, Chris Mason, Christstoo2, Cinik, Cjrother, Classic45, Cloveious, Coffee, Cogito-ergo-sum, Colinflies, Colonies Chris, CommonsDelinker, ConicProjection, Connormah, Cortomaltais, Cpatterson76, Cula, D-Rock, DMG413, Dale Arnett, Dale662, Darkcore, Darth Panda, David Levy, DavidWBrooks, Davish Krail, Gold Five, Dawn Bard, Dcandeto, Dcoetzee, Decumanus, Derek Ross, Dferrantino, Dieatwill1, Disneyfreak96, Dl2000, Don't Rain on my Burrito, Dove1950, Dpotter, Drooid, DuncanHill, Dunne409, Eileen R, EncMstr, Eneufeld, Enzo Aquarius, Ezeu, F.bendik, Facts707, Fat pig73, Fawcett5, Fire 55, Flyguy649, Formeruser0910, Fred Bauder, Frehley, G2bambino, Gaius Cornelius, Gauss, Geodyde, George415, Ghirlandajo, Gimlei, Glacier109, GorleyVMC, Gpvos, Graham87, GreenReaper, Greenshed, GregAsche, Gzornenplatz, HJKeats, Hairy Dude, Hdt83, Helli213, Hephaestos, Hillock65, Howardjp, Hugo999, Hunter1084, Imaginary heroes, Indefatigable, Infrogmation, Itai, Itschris, J.delanoy, J2rome, Jaccos, James500, JamesTeterenko, Jaraalbe, Java13690, Jdobbin, Jeff G., JenniferHeartsU, Jeronimo, Jfitzg, JillandJack, Jogloran, John Quiggin, John wesley, Johnyboy1997, Jon vs, Jonesy22, Josephprymak, Jusdafax, Kaare, Kablammo, Kaszkawal, Katr67, Kchishol1970, Keilana, Kelisi, KenWalker, Kevintoronto, Kittybrewster, Kmeister, Kmsiever, KnightRider, Kurieeto, Kzaral, Lambertman, Landroo, Langhorner, LeighvsOptimvsMaximvs, Leslie Mateus, Lesliedbruce, Lexicon, Light99, Lightmouse, Llywrch, Lockz, Looxix, Lupin, MJCdetroit, MONGO, Madchester, Madmagic, Marco Thomas, Master Jay, Mattbr, Matzthebomb, Maury Markowitz, Mav, Maximus Rex, Mb1000, Mdazey, Mechamind90, Member, Menchi, Mike Rosoft, Mindmatrix, Minesweeper, Mintguy, Mirv, Mistercow, Mitsukai, Mkdw, Modest Genius, Montrealais, Mrpoisson, Mrtrey99, Mscuthbert, Muchness, Murderbike, Mwanner, N00berson, Nakon, Nathaniel Christopher, NawlinWiki, Nerval, NorCalHistory, Olivier, P199, Paddu, Paul Drye, Pearle, Penguin020, Peruvianllama, Peterkingiron, Pfly, Pigman, Piledhigheranddeeper, Pinethicket, Polly, Ponder, Poor Yorick, QueenCake, Qutezuce, Qyd, R'n'B, Retired username, Retsopllib, Rich Farmbrough, Ringrust, Rlquall, Rmhermen, RobertG, Robocoder, Ronhjones, Rosemaryamey, Rosiestep, Roux, Ruzulo, RxS, Ryanamy83, Rymel, SDY, SFrank85, SYSS Mouse, Sam Blacketer, Sam Hocevar, Sardanaphalus, Sceptre, SchuminWeb, Scumoftheearth, Sdfisher, Secretlondon, Sfphotocraft, Sharabura, Shawn in Montreal, Shawnc, Shawtower76, Shay12345, ShelfSkewed, Singularity42, Sjö, Skookum1, Skyfaller, Speer320, Spinboy, Springnuts, Squilax, StanZegel, Steelbeard1, Stefanomione, Steve G, Stickguy, Stormbay, Svetulco1997, Svgalbertian, SweetNightmares, T. Mazzei, Tassedethe, Techietim, Telso, Tesi1700, The Thing That Should Not Be, The wub, Thegn, Themepark, Thickslab, TigerShark, Timc, Timurite, Toddst1, Toiyabe, Toonmon2005, TorontoStorm, Tran77, TruthbringerToronto, Tuxide, Udonknome, Urbanrenewal, Useight, Usgnus, VanessaCop, Vfp15, Vikramsidhu, VolatileChemical, Voyager, Wcreed88, Wehwalt, Wetman, Wikidea, Wikiuser100, Womble, Wsiegmund, Wtshymanski, Xenophrenic, YUL89YYZ, Zaintoum, Zaui, Zippymarmalade, 527 anonymous edits

Port Said *Source*: http://en.wikipedia.org/w/index.php?oldid=361452255 *Contributors*: *drew, Ahmad Soliman, Ahn J, AlbertHerring, Alex43223, Amikake3, Arab League, Ashashyou, Baros1, Bobo192, CRKingston, Canationalist, Chapultepec, Charlesdrakew, Clair4133, CrimeCentral, D6, Davepape, David de Cooman, Dawkeye, Dbachmann, DePiep, Dpm64, Dr. Blofeld, Dreadstar, Dryke, Ebsawy, Egyegy, Elmondo21st, Flyingbird, GDonato, Ghalas, Gilgamesh, Gmcase0, Golbez, Grim23, Grutness, Hede2000, Hmains, Huhsunqu, IZAK, Islamomt, Jaraalbe, Jason Recliner, Esq., Jennica, Kashwaa, Kitia, Lanternix, Lubos, MJCdetroit, Malcolma, Mario1952, Mark.murphy, Martpol, McTrixie, Mhsanabary9, Mmcannis, Mr Accountable, N2e, NawlinWiki, Nichalp, Nika 243, Olivier, Owen, Phevos87, Phil Boswell, Polylerus, Poor Yorick, Profoss, Psamtik 1, Radarino, Raul654, Rickard Vogelberg, Rockfang, Ronhjones, SchuminWeb, Skier Dude, Socal gal at heart, Station1, Taichi, Tbone762, TheEgyptian, TheParanoidOne, Vegaswikian, Wikid77, Wikimike, Wikix, Woohookitty, Ww2censor, Zazaban, 58 anonymous edits

Mumbai Fire Brigade *Source*: http://en.wikipedia.org/w/index.php?oldid=358243013 *Contributors*: Accesswilmington, D6, DAJF, Daysleeper47, Download, Gaurav, Howsa12, Iridescent, Jammy simpson, Kensplanet, LilHelpa, Malcolma, NawlinWiki, Ninetyone, Queenmomcat, Scottalter, Shyamsunder, Sniperz11, Suyogaerospace, Trafford09, Whpq, आशीष भटनागर, 10 anonymous edits

Karachi *Source*: http://en.wikipedia.org/w/index.php?oldid=363776261 *Contributors*: *Paul*, *drew, 1297, 15jan19932009, 1orbitx, 450w, 49danesway, 49oxen, 52xmax, 62.253.64.xxx, 7day, A Fantasy, A Nobody, A. B., AMZACCESS, ASOTMKX, AYousefzai, Aalahazrat, Aamerabbas, Aarandir, Aarsalankhalid, Aayan1, Aazai, Abasyn, AbdulQadir, Abstrakt, Acalamari, Achangeisasgoodasa, Adam.J.W.C., Adeel101, Aditya, Adnann, Advil, After Midnight, Agceltics, Agnte, Ahmed27, Ahn J, Ahoerstemeier, Ahsaninam, Ahsaniqbal 93, Ahsonkhan, Ahussains, Akaramatullah, Akarkera, Akchishti, Akhan1595, Alam82, Alansohn, AlasdairGreen27, AlbertR, Ali Rana, Ali Yaver, AliAmrohvi, Alijooan, Alirox, AlphaGamma1991, Altahafut, Amanatali.goher, Ambarish, Ambi92, Amorrow, Anarchangel, Angusmclellan, Anshuk, AntonioDsouza, AntonioMartin, Anupam, Anuraagvaidya, Anvar Samadzoda, Apexapexapex, Apokrif, Apparition11, Arcane0, Arhijazi, Arisa, Arjun01, ArsalanKhan, Arved, Arvind Iyengar, Asadabidi, Asifniz, Aslamt, Asmat Jamal, Ata Fida Aziz, Atitarev, AtticusX, Auric, Aursani, Aylahs, BD2412, BLADE, Badkhan, Baitulilm, Balcer, BananaFiend, Banksy1988, Barek, Bart133, Barticus88, Behemoth, Bellhalla, Bender235, Bennetto, Betterusername, Bhadani, BigHaz, Birdeditor, Blackeaglz, BobWoolmermate, Bobblewik, Bogdan, Bokhari76, BorgQueen, Brahui, Brainlara73, Bravo.0345, Brewcrewer, BrotherFlounder, Burgundavia, Byrial, C.Fred, CRKingston, CSWarren, Caiaffa, CalJW, Caltas, CambridgeBayWeather, Can't sleep, clown will eat me, Capricorn42, CardinalDan, Chachaji, Chantoke, Chaojoker, Chem1, ChiragPatnaik, Chowbok, Chris the speller, Cisguy, Citterio, Civilizationsschool, Classicrockfan42, Cleared as filed, Closedmouth, Cmdrjameson, Colipon, CommonsDelinker, ConcernedVancouverite, Conscious, Conversion script, Crotalus horridus, Cryptic, CryptoDerk, Cuffeparade, Current construction, Cwolfsheep, Cyberaks, DMacks, DaGizza, DancingMan, Daniel, Daniel Toth, Danishtaqvi, Danny, DaronDierkes, Datas, Ddgonzal, De Administrando Imperio, DeadEyeArrow, Deepak, Deeptrivia, Denking, Dennis Brown, DerHexer, Desayn, Dferrantino, DigiBullet, Discospinster, Disinterested, Djus, Dkc6004, Dlohcierekim, DocendoDiscimus, Docether, Download, Dpv, Dr hhq, Dr. Blofeld, DrAjitParkash, Drahsen, Drzeeshanali, DuKot, Dubbleup99, Dwaipayanc, ESkog, Earth, Edit khi, Editore99, Edward321, Effer, Ekabhishek, El C, Elockid, Emikofatima, Enric Naval, EoGuy, Eranb, Erebus Morgaine, Erwcmkiczfroipdeqw, Euchiasmus, Evotech, Excirial, Ezhiki, Ezra Wax, FARAZ KAYANI, Faisaltt, Faizan altaf, Farazars, Farazilu, Farhansher, Farhanweb, Farooq500, Fasail7, Fasibr, Fast track, Fconaway, Ferkelparade, Fhassan, Fieldday-sunday, Figgie123, FlyingToaster, Fr usc, Fratrep, Funandtrvl, Futurebird, G. Capo, GGreeneVa, Gabbe, Gadfium, Gamesmasterg9, Garwig, GeorgeOrr, Ghayyour, Giftlite, Gilded Lily, Gimboid13, Giraffedata, Gnowor, Godardesque, Godlord2, Gogo Dodo, Graymornings, Grayshi, Green Giant, GregorB, Grenavitar, Grillo, Guitarmankev1, Habibchem, Hadal, Haeleth, HappyInGeneral, Hariswaheed, Harryboyles, Haruth, Haslam1992, Hasnain.shahab, Hayatomermalik, HeBhagawan, Hellenica, HenkvD, HennessyC, Hillel, Himalayan Explorer, Hippietrail, Hirnee cheema, Hisroyalmajesty, Hmains, Hongooi, HotPakiBoy, Hu12, Hulleye, Humanyunkhann, Hydrox, IFaqeer, Ianbron, Iancaddy, Ibn-arabi, Idleguy, Igrek, Imran khan121, Indon, Interwal, Iquadri, Isenhand, Islamuslim, Islescape, Ismail.sultan, Ismaili101, Ithinkhelikesit, J Milburn, J.delanoy, J04n, JB82, JForget, JHCC, JLaTondre, JYi, Jackol, Jainrajat11, Jamalpanhwar, Jamatkhana, James Huddz, James500, JamesAM, Jamesontai, Jane McCann, Jangpk, Jari.mustonen, Jarry1250, Javit, Jdcooper, Jeff G., Jeff3000, Jeffersonghauri, Jimp, JoDonHo, JoanneB, Johan Magnus, JohnCD, Johnleemk, Jojit fb, Jordan 1972, Jose77, Jungli, KFP, KNewman, KO, Ka Faraq Gatri, Kamalfaridi, Kamranyosuf, Karachibreeze, Karachiite, Karachis-iceman, Karnarazdan, Kbdank71, Ken Gallager, Ketiltrout, Khalidkhoso, Khalido, Khirad, Khkhan, Khoikhoi, King of Hearts, Kingmobi, Kjetil r, Kjoonlee, Koavf, Kohlgill, Kohrook, Koolats, Kralizec!, Kshatriyaaz, Kslall8765, L-Tyrosine, Lahoreimage, LarryBH, LcawteHuggle, Le Grey, LeaveSleaves, Lestrade, Leuko, Lightmouse, Lilac Soul, Little Mountain 5, Loodog, Lostintherush, Lotoo, Luh-e, Luna Santin, Lupinoid, M maarij yousufi, M o b i, M-le-mot-dit, M.Imran, M.yousuf, M12390, M2k41, MC10, MECU, MER-C, MJCdetroit, MK8, Ma8ador, Maale27, Mackensen, Madhero88, Madman 0014, Madman445, Maijinsan, Mak2007, Malahat710, Mani shahrokni, Manizeh, Manra, MarSch, Marek69, Martin451, Marwatt, Masalai, MasoodABMQ, Masterrows, Mastoi01, Mathpianist93, Maulikvpatel, Maw503, Maximus Rex, Maya, Meco, Meemmohsin, Meherzaidi, Mehran Mangrio, Meraza, Michael Devore, Michellecrisp, Midgrid, MilborneOne, Mild Bill Hiccup, Milkmooney, Misaq Rabab, Mjs072, Mkamranb, Mm11, Mmansoor, Mohammad Akram R, Mohsin.Siddiqui, Monikahingorani, Mr Accountable, Mr. Wheely Guy, MrOllie, Mrabcx, Mralibhai, Mrizwankhan, Msferoz85, Muhammad Hamza, Muhammad ameen, Muhammadhani, Muhammdasif, Muhassan, Murad67, Murtaza Pakistani, Mustihussain, Mystifier-1, Nabeel82, Nadeemalikhan, Nappymonster, Nayvik, Nazimpak, Nazli, Neeraj nabar, Netdevil, Neutrality, Nichalp, NielsenGW, Nihiltres, Nikai, Nikkul, Nirav.maurya, Niteowlneils, Nixeagle, Njaelkies Lea, Nk, Nmkmathan, No badmashi, Noah Salzman, Nomi887, Noor Aalam, Noroton, Nsaa, Nucleus.one, Nv8200p, Oate238ca4ada42, Oatmeal batman, One2one, Oneiros, Oqureshi, Orestek, Ossmann, Pahari Sahib, Paine Ellsworth, Pajjar, Pakboy, Pakibabupeek, Pakipakpak, Pakistanidude, Paknur, Patoldanga'r Tenida, Paul August, Paul-L, Pepsidrinka, PeterCanthropus, Phil153, PhilKnight, Pk-user, Pk5abi, Plutonics, Pmj, Polaron, Postoak, Presidentman, QadeemMusalman, Quadell, Quarl, Quiddity, Ragib, Rama's Arrow, Rameezraja001, Ranveig, Rashid8928, Rashidr, Ratzd'mishukribo, Re-reconquista, Real Republic Ireland, Rebecca, RedWolf, Reedy, ReelExterminator, Reenem, Renesis, Rich Farmbrough, Richard D. LeCour, Rickterp, Rizwanayub, Rjwilmsi, RobDe68, Robert K S, Rorro, Runglirungliyot, Russavia, Rzafar, SDC, SElefant, SI4, Sa122, Saad awal, Saad.ghauri, SaadRajabali, Saadaslam, Sadafawan, Saffi2k7, Saga City, Saim1402, Saintrain, Sajidseo, Saki, Sam Hocevar, Sam8, Samar, Samuelsen, Samveen, Sansonic, Saq!b, Saqib.husain, Saqibrjpt, Sarah crane, Sardanaphalus, Sarosh, Sarz1986, Sashah666, Scetoaux, Scheibenzahl, Scholar2010, Schoolkaadon, Scope creep, SeanLegassick, Seano1, Ser Amantio di Nicolao, Sfacets, Shahroze, Shahrukh gazdar, Shakeelaamir, Shalom Yechiel, Shanes, Shanoo, Shariq, Sharpedge2, Shayanshaukat, Shehri cbe, Shigernafy, Shinas, Shirulashem, Shoejar, Siddhesh222, Siddiqui, Signalhead, Silversam, SimonP, Sirajahsan, Sirhanx2, Siroxo, Size J Battery, Skier Dude, Skoosh, Slides, SluggoOne, Sluzzelin, Smartpersonwithbrain, Smartspinhead, Smsarmad, Snagari, Sohail Shamim, Sohailstyle, Solar20, Someone the Person, Souvik.arko, Spasage, Splash, Sqamar, Srisez, Srose, Ssrashid, Stallions2010, Steave77, SteinbDJ, Steinsky, Sting14ray, Studerby, Suisui, Supersaiyan474, Supiwani, Suprah, Sweetgirl1965, Swerveut, Swifteagle43, Syed mohsin rizvee, SyedNaqvi90, Szhaider, T-borg, THEN WHO WAS PHONE?, THEunique, Tabletop, Taimoors, Taketa, Talhaahmed, Tanveer zeb, Tanzeelism, Tariqabjotu, Tatterfly, Tausifkhan, Tcncv, Techtrix, TedE, Test wiki00, Tfazlani, ThaGrind, The Wild West guy, The wub, Thue, Tim1988, TimBentley, Timir2, Tkhan, Tobby72, Todeskaefer,

TonyTheTiger, Toon05, Topperfalkon, Tpbradbury, Tractorkingsfan, Trakesht, Travellerswork, Triwbe, Trusilver, Turian, Turkeyleg779, Uchiah24, Ukexpat, Ulric1313, Umair ua, Umairkhatri, UnknownForEver, Unre4L, Unschool, Unyoyega, Uschfauq, Usman uk, Veinor, Vera.tetrix, Versageek, Victor, VikramRohra, Vpendse, Waqas.usman, Waqasahmedkain, Wasell, Wavelength, Welsh, Wereon, WhisperToMe, Wik, Wiki Wikardo, Wikid77, Wmahan, Woohookitty, Woudloper, Wtmitchell, Yadish, Yasiraq, Yasirniazkhan, YellowMonkey, Yerpo, Ynhockey, Yomangani, Zafthelawyer, Zaidiwaqas, Zaindy87, Zainubrazvi, Zeesh, Zeeshan8 8, Zeno of Elea, Zfr, Zinnmann, Zoicon5, Zoomzoom, Zubair71, Zuhair siddiqui, Zulfikkur, Zulfiqar.halari, Zuner Ahmed, ^demon, 영이, 2263 anonymous edits

Maharashtra *Source*: http://en.wikipedia.org/w/index.php?oldid=363876445 *Contributors*: 128.101.45.xxx, 18.25, 195.186.255.xxx, 200.191.188.xxx, 93gregsonl2, AMbroodEY, Aalapk, Abecedare, Abhijeet22, Abhijitsathe, Adam.J.W.C., AdjustShift, Advandana, Ae-a, Aeusoes1, Ajaxkroon, Akradecki, Akshat2, Alan Liefting, Ali, Ambuj.Saxena, Ameya.potadar, Amolsable, Anandks007, Anchitk, Ando228, AndrewHowse, Angusmclellan, Anilkirti, Anish7, Ankurem, Anthony, Anwar saadat, Armeria, Arteyu, Arunkrish, Arunsingh77, Arvindn, Ashurockstarboy, Ashycool, Asrghasrhiojadrhr, Asro 21, Atulkaluskar, Avenue, Avinashmane, Ayelkawar, BD2412, Babbage, Baronnet, Bidngrab, Bikehorn, Bkngp, Blah28948, Blurpeace, Bobby H. Heffley, BostonPunekar, Brighterorange, Byrial, CWesling, Caiaffa, CanadianLinuxUser, Canu44, Capricorn42, CarTick, CarolGray, Catdogpaw, Ccacsmss, Chanheigeorge, Charles Matthews, Chas zzz brown, Chekaz, Chinmay26r, ChiragPatnaik, Chowbok, Chrism, Citterio, ClaudineChionh, Closedmouth, Cometstyles, Commonman1, CommonsDelinker, Compwhiz, Conscious, Conversion script, Crazyvas, Crisspy, Csörföly D, DabMachine, Danianjan, Daniel Quinlan, Darklilac, David Kernow, David Shay, Deepak D'Souza, Dejvid, Den fjättrade ankan, DerHexer, Despardes7, Devanjedi, Devdattam, Dewan357, Dhoom, DigiBullet, Dilipb, Docu, Donotblkme, DrLeonP, Drilnoth, Drmies, DuKot, EGroup, Earl Andrew, Ecomantra, Edward, Ee00224, El C, Electromite, Fastily, Fconaway, Firsfron, Foochar, Fratrep, Fundamental metric tensor, Gadfium, Gaius Cornelius, Ganeshk, Gautambas, Geeteshgadkari, Gene Nygaard, GerardM, Ghakko, Gilgamesh, Giridhar, Gman124, Good Olfactory, Gppande, Gracenotes, Graham87, Grenavitar, Ground Zero, Gsingh, HalfShadow, Hdurina, Hemanshu, Hephaestos, Hintha, IXU79, IanBailey, Imc, Improv, ImpuMozhi, India Gate, India Rising, Info4all, Infogiventake, J.karan1105, JEH, JaGa, Jagged, Jahangard, Jambudweep, Jaxl, Jay, Jazzradio, Jeff G., Jeroen, John Hill, JohnI, JorgeGG, JorisvS, Jose Ramos, Joseph Solis in Australia, Jovianeye, Jusjih, Just H, Justarijit, Kamble0, Karan deshmukh9005, Katherine Shaw, Kathleen.wright5, Katimawan2005, Kaveri, Kedar Borhade, Kelly, Kensplanet, Kintetsubuffalo, Kitabparast, Kkm010, Knight45, Kostja, Kralizec!, Krishnak12, Ksumedh, Ktr101, Kummi, Kunalsupe, Kurykh, KuwarOnline, Kwamikagami, Kww, Lalit Jagannath, Laxmikant13, Layzyak, Leandrod, Lex Rex, Lightmouse, LilHelpa, Llywrch, Luka Jačov, Lysis, M3taphysical, MER-C, MPerel, Magicalsaumy, Mah exp, Maharashtraexpress, Mahawiki, Mahexp 2007, Mamta dhody, Mani1, Manishkoparkar, Mannoj, Manuel Anastácio, Marathi hero, Martarius, Mattisse, MaximvsDecimvs, Mayur.thakare, Mayuresh gapchup, Mayuresh.kathe, Meetkrp, MegaSloth, Meghanand, Mel Etitis, Metasur, Mhexp00000, Mhexp001, Morwen, MrOllie, Mrdrift, Mrtpolice, Mskadu, Mumbaicha, Mxsla, Nadkaarnee, Nat Krause, Natrajdr, Naveenbm, Navvis, Neurolysis, New Rock Star, Nichalp, Nihiltres, Nikkul, Nilheda, Ninadhardikar, Nishant12, Notinasnaid, Nposs, Nskamat, Obli, Obradovic Goran, Oh Snap, Ohnoitsjamie, Olivier, OneGuy, Orrorin, Ortcutt, PCHS-NJROTC, PRITAM9890093772, Paalappoo, Paddu, Paine Ellsworth, Paul-L, Pax:Vobiscum, Paxse, Pgk, Pharaoh of the Wizards, Pizzadeliveryboy, Planemad, Plrk, Poojakaul, Potomac123, Prabhak, Pradeepbhadikar, PrasadhBaapaat, Prashanthns, Praveengedam, Pritamgshah, Priyanath, Proofreader77, Radar signal, Radhey, RafaAzevedo, Rahul9890177973, Railwayweb, Rakeshkallurkar, Ranveig, Ratzd'mishukribo, RedWolf, Redtigerxyz, Reedy, Rich Farmbrough, Rick Block, Ricky81682, Rivatphil, Rjwilmsi, Rmckarthik, Rock2e, Rohitbd, Roux, Rowheat, Rpagnis, Rsmn, Rsrikanth05, Rushikeshshinde, SNIyer12, Sachin f, Sachinjoshi, Sahodaran, Saimdusan, Saitan.das, SalilSBudhe, Salilb, Sam.khawse, Sangharsh, Sanketholey, Sarangdixit, Saravask, Sarvabhaum, Saurabhkulkarni, Scanlan, Sceptre, Semposition2, Shanes, Shivap, Shivashree, Shj95, Shrivipha, Shyamsunder, Siddhant, Sidmoe, Siebrand, Sietse Snel, Signalhead, Siroxo, SivaKumar, Skarebo, Skarl the Drummer, Skokatay, SkyWalker, Snoyes, Soman, SpacemanSpiff, Spartiate, Spundun, Squids and Chips, Sribalajisociety, Srikeit, Stateofart, SuchiBhasin, Sue H. Ping, Sufficient Memory, Sundar, Suneetk, Super cyclist, Sushant gupta, Swapnilmandhalkar, T-borg, Tabletop, Teh tennisman, Tejas81, Template namespace initialisation script, Terissn, Thaurisil, The idiot, Thebigdig, Thegreyanomaly, Thesmothete, ThomasPhilip.jr, TimR, Tinttong, Tobias Conradi, Tobias Hoevekamp, Tom Radulovich, Trakesht, Tuncrypt, Turian, Unyoyega, Utkarsh sawale, Vaibhavmit, Vaikunda Raja, Vayu, Vbasrur, Vdeolali, Vedran12, Victor D, Vijayaditya, Vikripedia, Vikru2007, Vishal1976, Voyevoda, Vvp1001, Vvuppala, Water Fish, Wiccan Quagga, Wik, Woohookitty, Yash chauhan2, YellowMonkey, Zachorious, ~shuri, 919 anonymous edits

Image Sources, Licenses and Contributors

Image:Bombay-Docks-aftermath1.png *Source*: http://en.wikipedia.org/w/index.php?title=File:Bombay-Docks-aftermath1.png *License*: Public Domain *Contributors*: Nichalp, Roland zh

Image:SS Fort Stikine.png *Source*: http://en.wikipedia.org/w/index.php?title=File:SS_Fort_Stikine.png *License*: Public Domain *Contributors*: Korrigan, Nichalp, Roland zh, Sankalpdravid

Image:Bombay-Docks-aftermath3.png *Source*: http://en.wikipedia.org/w/index.php?title=File:Bombay-Docks-aftermath3.png *License*: Public Domain *Contributors*: Nichalp, Roland zh

Image:1944 Bombay harbour explosion propeller piece.jpg *Source*: http://en.wikipedia.org/w/index.php?title=File:1944_Bombay_harbour_explosion_propeller_piece.jpg *License*: Creative Commons Attribution-Sharealike 3.0 *Contributors*: User:Nichalp

File:Mumbai_Montage.jpg *Source*: http://en.wikipedia.org/w/index.php?title=File:Mumbai_Montage.jpg *License*: Creative Commons Attribution-Sharealike 3.0 *Contributors*: Cirt, IngerAlHaosului, Nikkul, Redtigerxyz, 2 anonymous edits

File: India Maharashtra location map.svg *Source*: http://en.wikipedia.org/w/index.php?title=File:India_Maharashtra_location_map.svg *License*: Creative Commons Attribution-Sharealike 3.0 *Contributors*: Maximilian Dörrbecker (Chumwa)

File:India Maharashtra locator map.svg *Source*: http://en.wikipedia.org/w/index.php?title=File:India_Maharashtra_locator_map.svg *License*: unknown *Contributors*: Abhijitsathe, Planemad

File:Red pog.svg *Source*: http://en.wikipedia.org/w/index.php?title=File:Red_pog.svg *License*: Public Domain *Contributors*: User:Andux

File:Flag of India.svg *Source*: http://en.wikipedia.org/w/index.php?title=File:Flag_of_India.svg *License*: Public Domain *Contributors*: User:SKopp

File:Loudspeaker.svg *Source*: http://en.wikipedia.org/w/index.php?title=File:Loudspeaker.svg *License*: Public Domain *Contributors*: Bayo, Gmaxwell, Husky, Iamunknown, Nethac DIU, Omegatron, Rocket000, 5 anonymous edits

File:Mumbadevi temple.jpg *Source*: http://en.wikipedia.org/w/index.php?title=File:Mumbadevi_temple.jpg *License*: Creative Commons Attribution-Sharealike 2.0 *Contributors*: Magiceye

File:Kanheri-stupa1.jpg *Source*: http://en.wikipedia.org/w/index.php?title=File:Kanheri-stupa1.jpg *License*: Creative Commons Attribution-Sharealike 2.5 *Contributors*: User:Nichalp

File:Hajiali.jpg *Source*: http://en.wikipedia.org/w/index.php?title=File:Hajiali.jpg *License*: Creative Commons Attribution-Sharealike 2.0 *Contributors*: Humayunn Peerzaada AKA HumFur from Mumbai, India

File:Ships in Bombay Harbour, 1731.jpg *Source*: http://en.wikipedia.org/w/index.php?title=File:Ships_in_Bombay_Harbour,_1731.jpg *License*: Public Domain *Contributors*: Unknown

Image:Hutatma Chowk.jpg *Source*: http://en.wikipedia.org/w/index.php?title=File:Hutatma_Chowk.jpg *License*: Creative Commons Attribution-Sharealike 2.0 *Contributors*: Elroy Serrao

File:Mumbaicitydistricts.png *Source*: http://en.wikipedia.org/w/index.php?title=File:Mumbaicitydistricts.png *License*: unknown *Contributors*: Bobo192, Stateofart, 3 anonymous edits

File:India mumbai temperature precipitation averages chart.svg *Source*: http://en.wikipedia.org/w/index.php?title=File:India_mumbai_temperature_precipitation_averages_chart.svg *License*: GNU Free Documentation License *Contributors*: Own work

File:BSE.jpg *Source*: http://en.wikipedia.org/w/index.php?title=File:BSE.jpg *License*: Creative Commons Attribution-Sharealike 2.0 *Contributors*: Uploaded to Flickr on August 18, 2007 by Flickr user anappaiah). Uploaded to the English Wikipedia by

File:Mumbai Skyline1.jpg *Source*: http://en.wikipedia.org/w/index.php?title=File:Mumbai_Skyline1.jpg *License*: Creative Commons Attribution-Sharealike 3.0 *Contributors*: User:Deepak

File:Highcourt.jpg *Source*: http://en.wikipedia.org/w/index.php?title=File:Highcourt.jpg *License*: Creative Commons Attribution-Sharealike 2.0 *Contributors*: Michael Siegel AKA michaelj-siegel

File:1st INC1885.jpg *Source*: http://en.wikipedia.org/w/index.php?title=File:1st_INC1885.jpg *License*: Public Domain *Contributors*: Unknown

File:BEST-Mumbai-Brtsnewaug08.jpg *Source*: http://en.wikipedia.org/w/index.php?title=File:BEST-Mumbai-Brtsnewaug08.jpg *License*: Creative Commons Attribution-Sharealike 3.0 *Contributors*: Another Indian Citizen

File:Mumbai Train Station.jpg *Source*: http://en.wikipedia.org/w/index.php?title=File:Mumbai_Train_Station.jpg *License*: Creative Commons Attribution-Sharealike 2.0 *Contributors*: FlickreviewR, Gryffindor, Indianhilbilly

File:Mumbai Airport.jpg *Source*: http://en.wikipedia.org/w/index.php?title=File:Mumbai_Airport.jpg *License*: Creative Commons Attribution-Sharealike 2.0 *Contributors*: Alex Graves from Lugano, Switzerland

File:Bombay Municipal Corporation.JPG *Source*: http://en.wikipedia.org/w/index.php?title=File:Bombay_Municipal_Corporation.JPG *License*: Creative Commons Attribution-Sharealike 3.0 *Contributors*: User:Vegpuff

File:Dharavi Slum in Mumbai.jpg *Source*: http://en.wikipedia.org/w/index.php?title=File:Dharavi_Slum_in_Mumbai.jpg *License*: Creative Commons Attribution-Sharealike 3.0 *Contributors*: Kounosu

File:Town-Hall,-Bombay.jpg *Source*: http://en.wikipedia.org/w/index.php?title=File:Town-Hall,-Bombay.jpg *License*: Creative Commons Attribution-Sharealike 2.5 *Contributors*: User:Nichalp

File:Elephanta Caves Trimurti.jpg *Source*: http://en.wikipedia.org/w/index.php?title=File:Elephanta_Caves_Trimurti.jpg *License*: Creative Commons Attribution 2.0 *Contributors*: Christian Haugen

File:Lalbaughcha raja.jpg *Source*: http://en.wikipedia.org/w/index.php?title=File:Lalbaughcha_raja.jpg *License*: Public Domain *Contributors*: User:Megaeclipse

File:Flag of the United Kingdom.svg *Source*: http://en.wikipedia.org/w/index.php?title=File:Flag_of_the_United_Kingdom.svg *License*: Public Domain *Contributors*: User:Zscout370

File:Flag of the United States.svg *Source*: http://en.wikipedia.org/w/index.php?title=File:Flag_of_the_United_States.svg *License*: Public Domain *Contributors*: User:Dbenbenn, User:Indolences, User:Jacobolus, User:Technion, User:Zscout370

File:Flag of Germany.svg *Source*: http://en.wikipedia.org/w/index.php?title=File:Flag_of_Germany.svg *License*: Public Domain *Contributors*: User:Pumbaa80

File:Flag of Russia.svg *Source*: http://en.wikipedia.org/w/index.php?title=File:Flag_of_Russia.svg *License*: Public Domain *Contributors*: AndriusG, Davepape, Dmitry Strotsev, Drieskamp, Enbéká, Fred J, Gleb Borisov, Herbythyme, Homo lupus, Kiensvay, Klemen Kocjancic, Kwj2772, Mattes, Maximaximax, Miyokan, Nightstallion, Ondřej Žváček, Pianist, Pumbaa80, Putnik, R-41, Radziun, Rainman, Reisio, Rfc1394, Rkt2312, Rocket000, Sasa Stefanovic, SeNeKa, SkyBon, Srtxg, Stianbh, Wikiborg, Winterheart, Zscout370, Zyido, ОйЛ, 34 anonymous edits

File:Flag of Japan.svg *Source*: http://en.wikipedia.org/w/index.php?title=File:Flag_of_Japan.svg *License*: Public Domain *Contributors*: Various

File:Bolywood.jpg *Source*: http://en.wikipedia.org/w/index.php?title=File:Bolywood.jpg *License*: Creative Commons Attribution-Sharealike 2.0 *Contributors*: Ekabhishek, FlickreviewR, Indianhilbilly, Martin H., Mattes, Ranveig, Varunbhandanker

File:Mumbai Clock Towerr.jpg *Source*: http://en.wikipedia.org/w/index.php?title=File:Mumbai_Clock_Towerr.jpg *License*: Creative Commons Attribution-Sharealike 3.0 *Contributors*: User:Nikkul

File:Brabourne.jpg *Source*: http://en.wikipedia.org/w/index.php?title=File:Brabourne.jpg *License*: Creative Commons Attribution-Sharealike 2.0 *Contributors*: ojas c at flickr; Original uploader was Nikkul at en.wikipedia

File:Prince_Rupert_from_Mount_Morse_Nov_6th_2005-DSC_0725-700w.jpg *Source*: http://en.wikipedia.org/w/index.php?title=File:Prince_Rupert_from_Mount_Morse_Nov_6th_2005-DSC_0725-700w.jpg *License*: Creative Commons Attribution-Sharealike 2.5 *Contributors*: Alex Batko

File:Prince Rupert-COA.png *Source*: http://en.wikipedia.org/w/index.php?title=File:Prince_Rupert-COA.png *License*: unknown *Contributors*: Qyd, RyanGerbil10

Image:Canada British Columbia location map.svg *Source*: http://en.wikipedia.org/w/index.php?title=File:Canada_British_Columbia_location_map.svg *License*: Creative Commons Attribution-Sharealike 3.0 *Contributors*: NordNordWest

Image:Red pog.svg *Source*: http://en.wikipedia.org/w/index.php?title=File:Red_pog.svg *License*: Public Domain *Contributors*: User:Andux

File:Flag of Canada.svg *Source*: http://en.wikipedia.org/w/index.php?title=File:Flag_of_Canada.svg *License*: Public Domain *Contributors*: User:E Pluribus Anthony, User:Mzajac

File:Flag of British Columbia.svg *Source*: http://en.wikipedia.org/w/index.php?title=File:Flag_of_British_Columbia.svg *License*: Public Domain *Contributors*: specification sheets by Christopher Southworth, -xfi-'s source code

Image:Prince Rupert-May 1910-LP984-29-1759-368-700w.jpg *Source*: http://en.wikipedia.org/w/index.php?title=File:Prince_Rupert-May_1910-LP984-29-1759-368-700w.jpg *License*: Creative Commons Attribution-Sharealike 2.5 *Contributors*: Original uploader was Alexbatko at en.wikipedia

File:Capitol Mall and theatre in Prince Rupert, British Columbia.jpg *Source*: http://en.wikipedia.org/w/index.php?title=File:Capitol_Mall_and_theatre_in_Prince_Rupert,_British_Columbia.jpg *License*: Creative Commons Attribution-Sharealike 3.0 *Contributors*: User:Wknight94

Image:Orthographic projection centred over Prince Rupert.png *Source*: http://en.wikipedia.org/w/index.php?title=File:Orthographic_projection_centred_over_Prince_Rupert.png *License*: GNU Free Documentation License *Contributors*: Original uploader was Geo Swan at en.wikipedia

Image:Prince Rupert and Vancouver on the BC Coast.png *Source*: http://en.wikipedia.org/w/index.php?title=File:Prince_Rupert_and_Vancouver_on_the_BC_Coast.png *License*: Creative Commons Attribution-Sharealike 3.0 *Contributors*: User:Geo Swan

File:City Hall, Prince Rupert, British Columbia.jpg *Source*: http://en.wikipedia.org/w/index.php?title=File:City_Hall,_Prince_Rupert,_British_Columbia.jpg *License*: Creative Commons Attribution-Sharealike 3.0 *Contributors*: User:Wknight94

File:Totem poles near City Hall, Prince Rupert, British Columbia.jpg *Source*: http://en.wikipedia.org/w/index.php?title=File:Totem_poles_near_City_Hall,_Prince_Rupert,_British_Columbia.jpg *License*: Creative Commons Attribution-Sharealike 3.0 *Contributors*: User:Wknight94

Image:Prince Rupert Harbour.png *Source*: http://en.wikipedia.org/w/index.php?title=File:Prince_Rupert_Harbour.png *License*: GNU Free Documentation License *Contributors*: Original uploader was Geo Swan at en.wikipedia

File:Sunken Gardens in Prince Rupert, British Columbia.jpg *Source*: http://en.wikipedia.org/w/index.php?title=File:Sunken_Gardens_in_Prince_Rupert,_British_Columbia.jpg *License*: Creative Commons Attribution-Sharealike 3.0 *Contributors*: User:Wknight94

File:HBC Logo.svg *Source*: http://en.wikipedia.org/w/index.php?title=File:HBC_Logo.svg *License*: unknown *Contributors*: Connormah, 1 anonymous edits

Image:Hudsons Bay Company Flag.svg *Source*: http://en.wikipedia.org/w/index.php?title=File:Hudsons_Bay_Company_Flag.svg *License*: Public Domain *Contributors*: Cloveious, Denelson83, Denisutku, Fry1989, Lexicon, Liftarn, Makaristos, Skeezix1000, 1 anonymous edits

Image:Wpdms ruperts land.jpg *Source*: http://en.wikipedia.org/w/index.php?title=File:Wpdms_ruperts_land.jpg *License*: GNU Free Documentation License *Contributors*: -xfi-, Pyramide

Image:Hudsons Bay-Logo-Old.JPG *Source*: http://en.wikipedia.org/w/index.php?title=File:Hudsons_Bay-Logo-Old.JPG *License*: Creative Commons Attribution-Sharealike 2.5 *Contributors*: User:Qyd

Image:HBC-coa.JPG *Source*: http://en.wikipedia.org/w/index.php?title=File:HBC-coa.JPG *License*: Creative Commons Attribution-Sharealike 2.5 *Contributors*: User:Qyd

Image:Hbc post Lake Winnipeg.jpg *Source*: http://en.wikipedia.org/w/index.php?title=File:Hbc_post_Lake_Winnipeg.jpg *License*: Public Domain *Contributors*: AEMoreira042281, Er Komandante, Jkelly, P199

Image:hudsonsbayco.jpg *Source*: http://en.wikipedia.org/w/index.php?title=File:Hudsonsbayco.jpg *License*: GNU Free Documentation License *Contributors*: GeorgHH, Jaranda, Nordelch, Skeezix1000, Themedpark

File:Flag Egy PortSaid.gif *Source*: http://en.wikipedia.org/w/index.php?title=File:Flag_Egy_PortSaid.gif *License*: GNU Free Documentation License *Contributors*: Original uploader was Ninane at nl.wikipedia

File:Suez Canal, Port Said - ISS 2.jpg *Source*: http://en.wikipedia.org/w/index.php?title=File:Suez_Canal,_Port_Said_-_ISS_2.jpg *License*: Public Domain *Contributors*: unnamed NASA astronaut

image:Egypt location map.svg *Source*: http://en.wikipedia.org/w/index.php?title=File:Egypt_location_map.svg *License*: GNU Free Documentation License *Contributors*: User:NordNordWest

File:Flag of Egypt.svg *Source*: http://en.wikipedia.org/w/index.php?title=File:Flag_of_Egypt.svg *License*: unknown *Contributors*: 16@r, Alnokta, Anime Addict AA, ArséniureDeGallium, BomBom, Denelson83, Dinsdagskind, Duesentrieb, F l a n k e r, Flad, Foroa, Fry1989, Herbythyme, Homo lupus, Iamunknown, Klemen Kocjancic, Kookaburra, Ludger1961, Lumijaguaari, Mattes, Moroboshi, Neq00, Nightstallion, OsamaK, Permjak, Reisio, Rimshot, Str4nd, ThomasPusch, Thyes, Vonvon, Wikiborg, Wikimedia is Communism, Überraschungsbilder, 27 anonymous edits

File:PortSaid Canal 1880.jpg *Source*: http://en.wikipedia.org/w/index.php?title=File:PortSaid_Canal_1880.jpg *License*: Public Domain *Contributors*: unknown / unbekannt

File:PortSaidEgypt2 byDanielCsorfoly.JPG *Source*: http://en.wikipedia.org/w/index.php?title=File:PortSaidEgypt2_byDanielCsorfoly.JPG *License*: Public Domain *Contributors*: Daniel Csörföly

Image:A unique house of Port Said.JPG *Source*: http://en.wikipedia.org/w/index.php?title=File:A_unique_house_of_Port_Said.JPG *License*: GNU Free Documentation License *Contributors*: Original uploader was TheEgyptian at en.wikipedia

Image:Port said eglise cophete.jpg *Source*: http://en.wikipedia.org/w/index.php?title=File:Port_said_eglise_cophete.jpg *License*: Public Domain *Contributors*: around 1910

File:Port Said.jpg *Source*: http://en.wikipedia.org/w/index.php?title=File:Port_Said.jpg *License*: Creative Commons Attribution-Sharealike 2.5 *Contributors*: Albinfo

File:Mumbai Fire Brigade Logo Enhanced.jpg *Source*: http://en.wikipedia.org/w/index.php?title=File:Mumbai_Fire_Brigade_Logo_Enhanced.jpg *License*: unknown *Contributors*: User:Suyogaerospace

File:MFBUniform.JPG *Source*: http://en.wikipedia.org/w/index.php?title=File:MFBUniform.JPG *License*: Creative Commons Attribution-Sharealike 3.0 *Contributors*: User:Suyogaerospace

File:MFBOfficers.JPG *Source*: http://en.wikipedia.org/w/index.php?title=File:MFBOfficers.JPG *License*: Creative Commons Attribution-Sharealike 3.0 *Contributors*: User:Suyogaerospace

File:MFBUniFM.JPG *Source*: http://en.wikipedia.org/w/index.php?title=File:MFBUniFM.JPG *License*: Creative Commons Attribution-Sharealike 3.0 *Contributors*: User:Suyogaerospace

File:MFD Bronto Sky lift.JPG *Source*: http://en.wikipedia.org/w/index.php?title=File:MFD_Bronto_Sky_lift.JPG *License*: Creative Commons Attribution 3.0 *Contributors*: User:Suyogaerospace

File:MFDFiretruckold.JPG *Source*: http://en.wikipedia.org/w/index.php?title=File:MFDFiretruckold.JPG *License*: Creative Commons Attribution 3.0 *Contributors*: User:Suyogaerospace

File:AntiquesFireEngineMFB.JPG *Source*: http://en.wikipedia.org/w/index.php?title=File:AntiquesFireEngineMFB.JPG *License*: Creative Commons Attribution-Sharealike 3.0 *Contributors*: User:Suyogaerospace

Image:MFBLadder.JPG *Source*: http://en.wikipedia.org/w/index.php?title=File:MFBLadder.JPG *License*: Creative Commons Attribution-Sharealike 3.0 *Contributors*: User:Suyogaerospace

Image:MFDAWPwithaFirefiter.JPG *Source*: http://en.wikipedia.org/w/index.php?title=File:MFDAWPwithaFirefiter.JPG *License*: Creative Commons Attribution-Sharealike 3.0 *Contributors*: User:Suyogaerospace

Image:Hydraulic Platformcloseup.JPG *Source*: http://en.wikipedia.org/w/index.php?title=File:Hydraulic_Platformcloseup.JPG *License*: Creative Commons Attribution-Sharealike 3.0 *Contributors*: User:Suyogaerospace

Image:TelescopinghydraulicplatformRemote.JPG *Source*: http://en.wikipedia.org/w/index.php?title=File:TelescopinghydraulicplatformRemote.JPG *License*: Creative Commons Attribution-Sharealike 3.0 *Contributors*: User:Suyogaerospace

File:KarachiMontage.jpg *Source*: http://en.wikipedia.org/w/index.php?title=File:KarachiMontage.jpg *License*: unknown *Contributors*: Tfazlani

File:Karachi logo.svg *Source*: http://en.wikipedia.org/w/index.php?title=File:Karachi_logo.svg *License*: unknown *Contributors*: Hydrox

File:Karachi Locator Sindh Pakistan.PNG *Source*: http://en.wikipedia.org/w/index.php?title=File:Karachi_Locator_Sindh_Pakistan.PNG *License*: GNU Free Documentation License *Contributors*: nomi887 (talk)

image:Pakistan location map.svg *Source*: http://en.wikipedia.org/w/index.php?title=File:Pakistan_location_map.svg *License*: Creative Commons Attribution-Sharealike 3.0 *Contributors*: User:NordNordWest

Image:429068485 5f192dcf9f o.jpg *Source*: http://en.wikipedia.org/w/index.php?title=File:429068485_5f192dcf9f_o.jpg *License*: unknown *Contributors*: http://www.flickr.com/photos/zainub/

Image:Napier Mole Bridge to Keamari -Karachi in 1900-.jpg *Source*: http://en.wikipedia.org/w/index.php?title=File:Napier_Mole_Bridge_to_Keamari_-Karachi_in_1900-.jpg *License*: Public Domain *Contributors*: Unknown

File:KhiairportWW2.jpg *Source*: http://en.wikipedia.org/w/index.php?title=File:KhiairportWW2.jpg *License*: Public Domain *Contributors*: Syed Umair Ali Hashmi

Image:STS087-715-70.JPG *Source*: http://en.wikipedia.org/w/index.php?title=File:STS087-715-70.JPG *License*: Public Domain *Contributors*: Darwinek, Fconaway, M.Imran, 1 anonymous edits

Image:Oyster Rocks.JPG *Source*: http://en.wikipedia.org/w/index.php?title=File:Oyster_Rocks.JPG *License*: Creative Commons Attribution 3.0 *Contributors*: Billmirza

File:Clifton.JPG *Source*: http://en.wikipedia.org/w/index.php?title=File:Clifton.JPG *License*: Public Domain *Contributors*: Docsid74, EugeneZelenko, Semnoz, Shinka

Image:Karachi population.svg *Source*: http://en.wikipedia.org/w/index.php?title=File:Karachi_population.svg *License*: Public Domain *Contributors*: User:Conscious

File:Korangi Road Karachi.jpg *Source*: http://en.wikipedia.org/w/index.php?title=File:Korangi_Road_Karachi.jpg *License*: unknown *Contributors*: Kilom691, Muhammad ameen, 1 anonymous edits

File:Ii chundrigar.jpg *Source*: http://en.wikipedia.org/w/index.php?title=File:Ii_chundrigar.jpg *License*: Public Domain *Contributors*: Kamranyosuf, 1 anonymous edits

Image:KarachiFinancial.jpg *Source*: http://en.wikipedia.org/w/index.php?title=File:KarachiFinancial.jpg *License*: Public Domain *Contributors*: Ianis, Nomi887, Shanes, Siddiqui, Sohailstyle, 3 anonymous edits

File:Mcbtower123.jpg *Source*: http://en.wikipedia.org/w/index.php?title=File:Mcbtower123.jpg *License*: Public Domain *Contributors*: Meemmohsin, Sdrtirs, 1 anonymous edits

Image:Mohatta Palace Karachi 2.jpg *Source*: http://en.wikipedia.org/w/index.php?title=File:Mohatta_Palace_Karachi_2.jpg *License*: Public Domain *Contributors*: User Swerveut on en.wikipedia

Image:Khi National Museum.jpg *Source*: http://en.wikipedia.org/w/index.php?title=File:Khi_National_Museum.jpg *License*: Public Domain *Contributors*: User:Shahid1024

File:Frere hall Karachi.jpg *Source*: http://en.wikipedia.org/w/index.php?title=File:Frere_hall_Karachi.jpg *License*: Public Domain *Contributors*: Kamranyosuf

file:Nat Std01.JPG *Source*: http://en.wikipedia.org/w/index.php?title=File:Nat_Std01.JPG *License*: Creative Commons Attribution-Sharealike 3.0 *Contributors*: User:Nomi887

File:Civic centre Karachi.jpeg *Source*: http://en.wikipedia.org/w/index.php?title=File:Civic_centre_Karachi.jpeg *License*: Creative Commons Attribution-Sharealike 3.0 *Contributors*: User:Nomi887

File:Karachi admin.PNG *Source*: http://en.wikipedia.org/w/index.php?title=File:Karachi_admin.PNG *License*: GNU Free Documentation License *Contributors*: User:Nomi887

File:PNEC.jpg *Source*: http://en.wikipedia.org/w/index.php?title=File:PNEC.jpg *License*: Public Domain *Contributors*: Saadumer

File:Lyari expressway.png *Source*: http://en.wikipedia.org/w/index.php?title=File:Lyari_expressway.png *License*: Creative Commons Attribution-Sharealike 3.0 *Contributors*: User:Nomi887

Image:Karachi Jinnah Airport.jpg *Source*: http://en.wikipedia.org/w/index.php?title=File:Karachi_Jinnah_Airport.jpg *License*: Public Domain *Contributors*: Swerveut, 5 anonymous edits

Image:CNG Bus Karachi.jpg *Source*: http://en.wikipedia.org/w/index.php?title=File:CNG_Bus_Karachi.jpg *License*: Creative Commons Attribution-Sharealike 3.0 *Contributors*: User:Aayan1

File:Flag of Mauritius.svg *Source*: http://en.wikipedia.org/w/index.php?title=File:Flag_of_Mauritius.svg *License*: Public Domain *Contributors*: User:Gabbe, User:SKopp

File:Flag of the People's Republic of China.svg *Source*: http://en.wikipedia.org/w/index.php?title=File:Flag_of_the_People's_Republic_of_China.svg *License*: Public Domain *Contributors*: User:Denelson83, User:SKopp, User:Shizhao, User:Zscout370

File:Flag of Hong Kong.svg *Source*: http://en.wikipedia.org/w/index.php?title=File:Flag_of_Hong_Kong.svg *License*: Public Domain *Contributors*: ButterStick, Dbenbenn, Jackl, Kibinsky, Kookaburra, Ludger1961, Mattes, Neq00, Nightstallion, Runningfridgesrule, Shinjiman, SkyBon, ThomasPusch, VAIO HK, Vzb83, Zscout370, 7 anonymous edits

File:Flag of Saudi Arabia.svg *Source*: http://en.wikipedia.org/w/index.php?title=File:Flag_of_Saudi_Arabia.svg *License*: Public Domain *Contributors*: some flag-maker

File:Flag of Uzbekistan.svg *Source*: http://en.wikipedia.org/w/index.php?title=File:Flag_of_Uzbekistan.svg *License*: Public Domain *Contributors*: User:Zscout370

File:Flag of Turkey.svg *Source*: http://en.wikipedia.org/w/index.php?title=File:Flag_of_Turkey.svg *License*: Public Domain *Contributors*: User:Dbenbenn

File:Flag of Lebanon.svg *Source*: http://en.wikipedia.org/w/index.php?title=File:Flag_of_Lebanon.svg *License*: Public Domain *Contributors*: Traced based on the CIA World Factbook with some modification done to the colours based on information at Vexilla mundi.

File:Flag of Bahrain.svg *Source*: http://en.wikipedia.org/w/index.php?title=File:Flag_of_Bahrain.svg *License*: Public Domain *Contributors*: User:SKopp, User:Zscout370

File:Flag of Kosovo.svg *Source*: http://en.wikipedia.org/w/index.php?title=File:Flag_of_Kosovo.svg *License*: GNU Free Documentation License *Contributors*: User:Cradel, User:Ningyou

File:Flag of Malaysia.svg *Source*: http://en.wikipedia.org/w/index.php?title=File:Flag_of_Malaysia.svg *License*: Public Domain *Contributors*: User:SKopp

Image:Karachi Cinema.jpg *Source*: http://en.wikipedia.org/w/index.php?title=File:Karachi_Cinema.jpg *License*: Creative Commons Attribution-Sharealike 1.0 *Contributors*: Fast track, Sfan00 IMG, 4 anonymous edits

Image:Market123.jpg *Source*: http://en.wikipedia.org/w/index.php?title=File:Market123.jpg *License*: Public Domain *Contributors*: Meemmohsin

Image:Korangi Road Karachi.jpg *Source*: http://en.wikipedia.org/w/index.php?title=File:Korangi_Road_Karachi.jpg *License*: unknown *Contributors*: Kilom691, Muhammad ameen, 1 anonymous edits

Image:Chaukundi1.JPG *Source*: http://en.wikipedia.org/w/index.php?title=File:Chaukundi1.JPG *License*: Creative Commons Attribution-Sharealike 2.5 *Contributors*: Original uploader was Mwaqas at en.wikipedia

Image:Kanab beach.JPG *Source*: http://en.wikipedia.org/w/index.php?title=File:Kanab_beach.JPG *License*: Public Domain *Contributors*: Zuhair siddiqui

Image:Karachi04.jpg *Source*: http://en.wikipedia.org/w/index.php?title=File:Karachi04.jpg *License*: Public Domain *Contributors*: Idleguy

Image:IICROAD.jpg *Source*: http://en.wikipedia.org/w/index.php?title=File:IICROAD.jpg *License*: Public Domain *Contributors*: Myself

Image:Mohatta Palace.jpg *Source*: http://en.wikipedia.org/w/index.php?title=File:Mohatta_Palace.jpg *License*: Public Domain *Contributors*: User:Shahid1024

Image:Karachi St. Patricks Cathedral.jpg *Source*: http://en.wikipedia.org/w/index.php?title=File:Karachi_St._Patricks_Cathedral.jpg *License*: Public Domain *Contributors*: XalD

File:Kemari Boat Basin @ Karachi.jpg *Source*: http://en.wikipedia.org/w/index.php?title=File:Kemari_Boat_Basin_@_Karachi.jpg *License*: Creative Commons Attribution 2.0 *Contributors*: Faisal Saeed

Image:FishingshipsatKarachiHarbour.JPG *Source*: http://en.wikipedia.org/w/index.php?title=File:FishingshipsatKarachiHarbour.JPG *License*: GNU Free Documentation License *Contributors*: Shayanshaukat (talk). Original uploader was Shayanshaukat at en.wikipedia. Later version(s) were uploaded by Skier Dude at en.wikipedia.

Image:CNBC Pakistan HQ at night.jpg *Source*: http://en.wikipedia.org/w/index.php?title=File:CNBC_Pakistan_HQ_at_night.jpg *License*: Creative Commons Attribution 2.0 *Contributors*: FlickreviewR, KevinAction, Zaccarias

File:Maharashtra Highlights.jpg *Source*: http://en.wikipedia.org/w/index.php?title=File:Maharashtra_Highlights.jpg *License*: Creative Commons Attribution 2.5 *Contributors*: User:Abhijitsathe

File: India_Maharashtra_locator_map.svg *Source*: http://en.wikipedia.org/w/index.php?title=File:India_Maharashtra_locator_map.svg *License*: unknown *Contributors*: Abhijitsathe, Planemad

File:Maharashtraseal.png *Source*: http://en.wikipedia.org/w/index.php?title=File:Maharashtraseal.png *License*: Public Domain *Contributors*: December21st2012Freak, Dn9ahx, Faizhaider, Ganeshk, Nichalp, Stannered, Stateofart, 1 anonymous edits

Image:Indischer Maler des 6. Jahrhunderts 001.jpg *Source*: http://en.wikipedia.org/w/index.php?title=File:Indischer_Maler_des_6._Jahrhunderts_001.jpg *License*: Public Domain *Contributors*: Gryffindor, Imz, Jastrow, 1 anonymous edits

Image:Marathas.GIF *Source*: http://en.wikipedia.org/w/index.php?title=File:Marathas.GIF *License*: Public Domain *Contributors*: Abhishekjoshi, Electionworld, Masur, Sankalpdravid

Image:Shivaji.jpg *Source*: http://en.wikipedia.org/w/index.php?title=File:Shivaji.jpg *License*: unknown *Contributors*: User:BotMultichillT

File:B. G. Tilak.gif *Source*: http://en.wikipedia.org/w/index.php?title=File:B._G._Tilak.gif *License*: Public Domain *Contributors*: unknown

Image:Pune India .jpg *Source*: http://en.wikipedia.org/w/index.php?title=File:Pune_India_.jpg *License*: Creative Commons Attribution-Sharealike 2.0 *Contributors*: Sunder Iyer

Image:Mahad Maharashtra.jpg *Source*: http://en.wikipedia.org/w/index.php?title=File:Mahad_Maharashtra.jpg *License*: Creative Commons Attribution-Sharealike 2.0 *Contributors*: Belasd, FlickreviewR, Indianhilbilly

Image:India Animals.jpg *Source*: http://en.wikipedia.org/w/index.php?title=File:India_Animals.jpg *License*: Creative Commons Attribution-Sharealike 2.0 *Contributors*: supersujit

Image:Mumbai skyline.jpg *Source*: http://en.wikipedia.org/w/index.php?title=File:Mumbai_skyline.jpg *License*: Creative Commons Attribution-Sharealike 2.0 *Contributors*: Amol.Gaitonde, Arria Belli, Belasd, FlickreviewR, Indianhilbilly, J o, Roland zh

Image:Mh elections.jpg *Source*: http://en.wikipedia.org/w/index.php?title=File:Mh_elections.jpg *License*: Creative Commons Attribution-Sharealike 3.0 *Contributors*: User:Fundamental metric tensor

Image:Highcourt.jpg *Source*: http://en.wikipedia.org/w/index.php?title=File:Highcourt.jpg *License*: Creative Commons Attribution-Sharealike 2.0 *Contributors*: Michael Siegel AKA michaelj-siegel

File:Clock Tower Mumbai University.jpg *Source*: http://en.wikipedia.org/w/index.php?title=File:Clock_Tower_Mumbai_University.jpg *License*: Creative Commons Attribution 2.0 *Contributors*: Steve Evans

Image:IITB Main Building.jpg *Source*: http://en.wikipedia.org/w/index.php?title=File:IITB_Main_Building.jpg *License*: GNU Free Documentation License *Contributors*: Ambuj.Saxena, Roland zh, 1 anonymous edits

File:Gsb.jpg *Source*: http://en.wikipedia.org/w/index.php?title=File:Gsb.jpg *License*: Public Domain *Contributors*: User:Hesvjti4

Image:Maharashtra Districts.png *Source*: http://en.wikipedia.org/w/index.php?title=File:Maharashtra_Districts.png *License*: Creative Commons Attribution-Sharealike 3.0 *Contributors*: User:Abhijitsathe

Image:Mumbai Skyline1.jpg *Source*: http://en.wikipedia.org/w/index.php?title=File:Mumbai_Skyline1.jpg *License*: Creative Commons Attribution-Sharealike 3.0 *Contributors*: User:Deepak

Image:Pune ShaniwarWada DelhiGate.jpg *Source*: http://en.wikipedia.org/w/index.php?title=File:Pune_ShaniwarWada_DelhiGate.jpg *License*: Creative Commons Attribution 2.5 *Contributors*: User:QuartierLatin1968

Image:Zero mile nagpur.jpg *Source*: http://en.wikipedia.org/w/index.php?title=File:Zero_mile_nagpur.jpg *License*: GNU Free Documentation License *Contributors*: nagpur.nic.in

Image:Nashik.jpg *Source*: http://en.wikipedia.org/w/index.php?title=File:Nashik.jpg *License*: Public Domain *Contributors*: User:Ashurockstarboy

Image:Aurangabad .jpg *Source*: http://en.wikipedia.org/w/index.php?title=File:Aurangabad_.jpg *License*: Creative Commons Attribution-Sharealike 2.0 *Contributors*: Gaurhav H. Atri

Image:Mumbai Airport.jpg *Source*: http://en.wikipedia.org/w/index.php?title=File:Mumbai_Airport.jpg *License*: Creative Commons Attribution-Sharealike 2.0 *Contributors*: Alex Graves from Lugano, Switzerland

Image:Mumbai Train Station.jpg *Source*: http://en.wikipedia.org/w/index.php?title=File:Mumbai_Train_Station.jpg *License*: Creative Commons Attribution-Sharealike 2.0 *Contributors*: FlickreviewR, Gryffindor, Indianhilbilly

Image:Kailasha temple at ellora.JPG *Source*: http://en.wikipedia.org/w/index.php?title=File:Kailasha_temple_at_ellora.JPG *License*: Creative Commons Attribution-Sharealike 2.5 *Contributors*: User:Pratheepps

Image:Ajanta (63).jpg *Source*: http://en.wikipedia.org/w/index.php?title=File:Ajanta_(63).jpg *License*: Creative Commons Attribution 2.5 *Contributors*: Belasd, Fagairolles 34, Nichalp, Soman, 1 anonymous edits

Image:India-Elephanta-Outside.jpg *Source*: http://en.wikipedia.org/w/index.php?title=File:India-Elephanta-Outside.jpg *License*: Public Domain *Contributors*: Gryffindor, Malo, ~shuri

Image:Bolywood.jpg *Source*: http://en.wikipedia.org/w/index.php?title=File:Bolywood.jpg *License*: Creative Commons Attribution-Sharealike 2.0 *Contributors*: Ekabhishek, FlickreviewR, Indianhilbilly, Martin H., Mattes, Ranveig, Varunbhandanker

Image:Dypatil 01.jpg *Source*: http://en.wikipedia.org/w/index.php?title=File:Dypatil_01.jpg *License*: GNU Free Documentation License *Contributors*: Jaxer

License

CPSIA information can be obtained at www.ICGtesting.com
Printed in the USA
LVOW050800260612

287678LV00003B/32/P